A Finite Element Primer for Beginners

Tarek I. Zohdi

A Finite Element Primer for Beginners

The Basics

Second Edition

Springer

Tarek I. Zohdi
University of California
Berkeley, CA
USA

ISBN 978-3-319-70427-2 ISBN 978-3-319-70428-9 (eBook)
https://doi.org/10.1007/978-3-319-70428-9

Library of Congress Control Number: 2014946399

Printed on acid-free paper

This Springer imprint is published by Springer Nature
The registered company is Springer International Publishing AG
The registered company address is: Gewerbestrasse 11, 6330 Cham, Switzerland

To my patient and loving wife, Britta Schönfelder.

Preface

The purpose of this primer is to provide the basics of the finite element method, primarily illustrated through a classical model problem, *linearized elasticity*. The topics covered are:

- Weighted residual methods and Galerkin's approximations,
- A model problem for one-dimensional linear elastostatics,
- Weak formulations in one dimension,
- Minimum principles in one dimension,
- Error estimation in one dimension,
- Construction of finite element basis functions in one dimension,
- Gaussian quadrature,
- Iterative solvers and element-by-element data structures,
- A model problem for three-dimensional linear elastostatics,
- Weak formulations in three dimensions,
- Basic rules for element construction in three dimensions,
- Assembly of the system and solution schemes,
- An introduction to time-dependent problems and
- An introduction to rapid computation based on domain decomposition and basic parallel processing.

The approach is to introduce the basic concepts first in one dimension and then move on to three dimensions. A relatively informal style is adopted. This primer is intended to be a "starting point," which can be later augmented by the large array of rigorous, detailed books in the area of finite element analysis. Through teaching finite element classes for a number of years at UC Berkeley, my experience has been that the fundamental weaknesses in prerequisite mathematics, such as vector calculus, linear algebra, and basic mechanics, exemplified by linearized elasticity, cause conceptual problems that impede the understanding of the finite element method. Thus, appendices on these topics have been included. Finally, I am certain that, despite painstaking efforts, there remain errors of one sort or another. Therefore, I would be grateful if readers who find such flaws would contact me at *zohdi@berkeley.edu.*

Berkeley, USA
September 2017

Tarek I. Zohdi

Contents

List of Figures

Weighted Residuals and Galerkin's Method for a Generic 1D Problem

1

1.1 Introduction: Weighted Residual Methods

Let us start by considering a simple one-dimensional differential equation, written in abstract form

$$A(u) = f, \tag{1.1}$$

where, for example, $A(u) = \frac{d}{dx}\left(A_1 \frac{du}{dx}\right) + A_2 u$, $A_1 = A_1(x)$ and $A_2 = A_2(x)$. Let us choose an approximate solution of the form

$$u^N = \sum_{i=1}^{N} a_i \phi_i(x), \tag{1.2}$$

where the ϕ_i's are approximation functions, and the a_i's are unknown constants that we will determine. Substituting the approximation leads to a "left over" amount called the residual:

$$r^N(x) = A(u^N) - f. \tag{1.3}$$

If we assume that the ϕ's are given, we would like to choose the a_i's to minimize the residual in an appropriate norm, denoted $||r||$. A primary question is which norm should be chosen to measure the solution and to determine its quality. Obviously, if the true solution is smooth enough to have pointwise solutions, and if we could take enough ϕ-functions, we could probably match the solution at every value of x. However, as we shall see, this would be prohibitively computationally expensive to solve. Thus, we usually settle for a less stringent measure, for example a spatially averaged measure of solution quality. This is not a trivial point, and we will formalize the exact choice of the appropriate metric (a norm) momentarily. Let us pick an obvious choice

$$\Pi(r^N) \stackrel{\text{def}}{=} ||r^N||^2 \stackrel{\text{def}}{=} \int_0^L (r^N(x))^2\, dx. \tag{1.4}$$

T. I. Zohdi, *A Finite Element Primer for Beginners*, The Basics,
https://doi.org/10.1007/978-3-319-70428-9_1

Taking the derivative with respect to each a_i, and setting it to zero, we obtain for $i = 1, 2, ...N$

$$\frac{\partial \Pi}{\partial a_i} = \int_0^L 2r^N \frac{\partial r^N}{\partial a_i}\, dx = 0. \tag{1.5}$$

This leads to N equations and N unknowns. This method is called the "Method of Least Squares." Another approach is to force the residual to be zero at a discrete number of locations, $i = 1, 2, ...N$

$$r^N(x_i) = 0, \tag{1.6}$$

which can also be written as

$$\int_0^L r^N(x)\delta(x - x_i)\, dx = 0, \tag{1.7}$$

where $\delta(x)$ is the Dirac Functional.[1] This approach is sometimes referred to as the "Collocation Method." Notice that each method has the form

$$\int_0^L r^N(x)w(x)\, dx = 0, \tag{1.9}$$

where $w(x)$ is some "weight." A general name for these methods is the "Method of Weighted Residuals."

1.2 Galerkin's Method

Of all of the weighted residual methods used in the scientific community, one particular method, the Galerkin method, is by far the most widely used and has been shown to deliver the most accurate solutions on a wide variety of problems. We now explain the basic construction. Consider the true solution, approximate solution and the error, related through

$$u - u^N = e^N \Rightarrow u = u^N + e^N. \tag{1.10}$$

As a helpful mnemonic, now consider them as vectors (Fig. 1.1). Clearly, the error (e^N) is the smallest when e^N is orthogonal to u^N. The problem is that the error

[1] Recall, the Dirac Functional is defined via

$$\int_0^L \delta(x - x_i) f(x)\, dx = f(x_i). \tag{1.8}$$

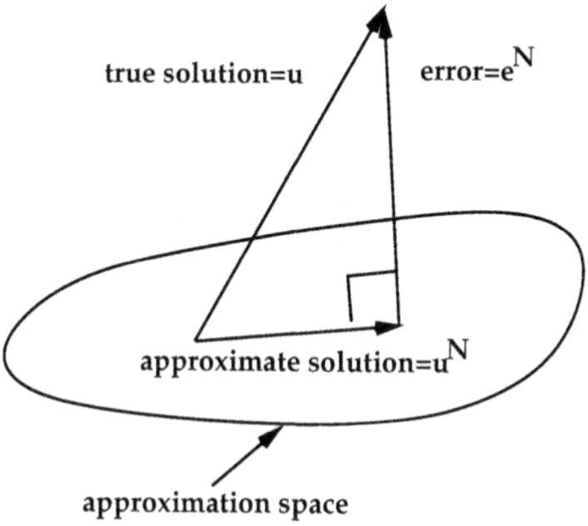

Fig. 1.1 Orthogonality of the approximation error

$e^N = u - u^N$ is unknown. However, the "next best thing," *the residual, is known.*[2] This motivates Galerkin's idea, namely to force u^N and r^N to be orthogonal. Mathematically, this can be expressed as

$$\int_0^L r^N(x)u^N(x)\,dx = \int_0^L r^N(x)\sum_{i=1}^{N} a_i\phi_i\,dx = 0. \tag{1.11}$$

However, this only gives one equation. Thus, we enforce this for each of the individual approximation functions, which collectively form the space of approximations comprising u^N,

$$\int_0^L r^N(x)a_i\phi_i(x)\,dx = a_i\int_0^L r^N(x)\phi_i(x)\,dx = 0 \Rightarrow \int_0^L r^N(x)\phi_i(x)\,dx = 0. \tag{1.12}$$

This leads to N equations and N unknowns, in order to solve for the a_i's. It is the usual practice in Galerkin's method to use approximation functions that are so-called kinematically admissible, which we define as functions that satisfy the so-called Dirichlet boundary conditions on u a priori.[3] Kinematically admissible functions do not have to satisfy boundary conditions that involve derivatives of the solution (u) beforehand.

[2]Although the error and residual are not the same, we note that when the residual is zero, the error is zero.

[3]The use of the phrase "kinematically admissible" comes from the fact that in early applications, the solution variable of interest was the displacement (u).

1.3 An Overall Framework

The basic "recipe" for the Galerkin process is as follows:

- Step 1: Compute the residual: $A(u^N) - f = r^N(x)$.
- Step 2: Force the residual to be orthogonal to each of the approximation functions: $\int_0^L r^N(x)\phi_i(x)\,dx = 0, i = 1, 2, 3 \ldots N$.
- Step 3: Solve the set of coupled equations. The equations will be linear if the differential equation is linear, and nonlinear if the differential equation is nonlinear.

The primarily problem with such a general framework is that it provides no systematic way of choosing the approximation functions. The basic finite element method has been designed to embellish and extend the fundamental Galerkin method by constructing ϕ_i's in order to deal with such issues. In particular:

- It is based upon Galerkin's method.
- It is computationally systematic and efficient.
- It is based on reformulations of the differential equations that remove the problems of restrictive differentiability requirements.

The approach that we will follow in this monograph is to introduce the basic concepts first in one dimension. We then present three-dimensional formulations, which extend naturally from one-dimensional formulations.

Remark: Two Appendices containing some essential background information on vector calculus, linear algebra, and basic mechanics, exemplified by linearized elasticity, are provided. Linearized elasticity will serve as our model problem in the chapters that follow.

A Model Problem: 1D Elastostatics

2

2.1 Introduction: A Model Problem

In most problems of mathematical physics the true solutions are nonsmooth; i.e., they are not continuously differentiable. *Thus, we cannot immediately apply a Galerkin approach.* For example in the equation of static mechanical equilibrium[1]

$$\nabla \cdot \boldsymbol{\sigma} + \boldsymbol{f} = \boldsymbol{0}, \tag{2.1}$$

there is an implicit requirement that the stress, $\boldsymbol{\sigma}$, is differentiable in the classical sense. Virtually the same mathematical structure form holds for other partial differential equations of mathematical physics describing diffusion, heat conduction, etc. *In many applications, differentiability is too strong a requirement, and in many locations it does not hold (the solution "jumps").* Therefore, when solving such problems we have two options: (1) enforcement of solution jump conditions at all of these locations (often they are not even known a priori) or (2) weak formulations (weakening the regularity requirements). Weak forms, which are designed to accommodate irregular data and solutions, are usually preferred. *Numerical techniques employing weak forms, such as the finite element method, have been developed with the essential property that whenever a smooth classical solution exists, it is also a solution to the weak form problem.* Therefore, we lose nothing by reformulating a problem in a more general way, by weakening the a priori smoothness requirements of the solution.

In the following few chapters, we shall initially consider a one-dimensional structure which occupies an open bounded domain in $\Omega \in I\!R$, with boundary $\partial\Omega$. The boundary consists of Γ_u on which the displacements (u), or any other primal variable (temperature in heat conduction applications, concentration in diffusion

[1]Here $\boldsymbol{f}$ are the body forces.

T. I. Zohdi, *A Finite Element Primer for Beginners*, The Basics,
https://doi.org/10.1007/978-3-319-70428-9_2

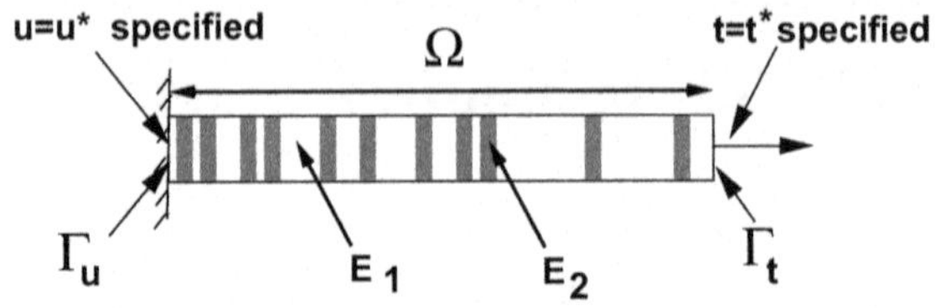

Fig. 2.1 A one-dimensional body

applications, etc. (see Appendix B)), are prescribed and a part Γ_t on which tractions ($t \stackrel{\text{def}}{=} \sigma n$, n being the outward normal) are prescribed ($t = t^*$, Fig. 2.1). We now focus on weak forms of a one-dimensional version of Eq. 2.1

$$\frac{d\sigma}{dx} + f = 0, \qquad (\sigma = E\frac{du}{dx}), \tag{2.2}$$

where $E = E(x)$ is a spatially varying coefficient (Fig. 2.1). Thereafter, we will discuss three-dimensional problems.

2.2 Weak Formulations in One Dimension

To derive a direct weak formulation for a body, we take Eq. 2.2 (denoted the strong form), form a product with an arbitrary smooth scalar-valued function ν, and integrate over the body

$$\int_\Omega (\frac{d\sigma}{dx} + f)\nu\, dx = \int_\Omega r\nu\, dx = 0, \tag{2.3}$$

where r is the residual. We call ν a "test" function. If we were to add a condition that we do this for all ($\stackrel{\text{def}}{=} \forall$) possible "test" functions then

$$\int_\Omega (\frac{d\sigma}{dx} + f)\nu\, dx = \int_\Omega r\nu\, dx = 0\, \forall \nu, \tag{2.4}$$

implies $r(x) = 0$. Therefore, if every possible test function were considered, then $r = \frac{d\sigma}{dx} + f = 0$ on any finite region in (Ω). Consequently, the weak and strong statements would be equivalent, provided the true solution is smooth enough to have a strong solution. Clearly, r can never be nonzero over any finite region in the body, because the test function will "find" them (Fig. 2.2). Using the product rule of differentiation on $\sigma\nu$ yields

$$\frac{d}{dx}(\sigma\nu) = (\frac{d\sigma}{dx})\nu + \sigma\frac{d\nu}{dx} \tag{2.5}$$

which leads to, $\forall\nu$

$$\int_\Omega (\frac{d}{dx}(\sigma\nu) - \sigma\frac{d\nu}{dx})\, dx + \int_\Omega f\nu\, dx = 0, \tag{2.6}$$

Fig. 2.2 Test functions actions on residuals

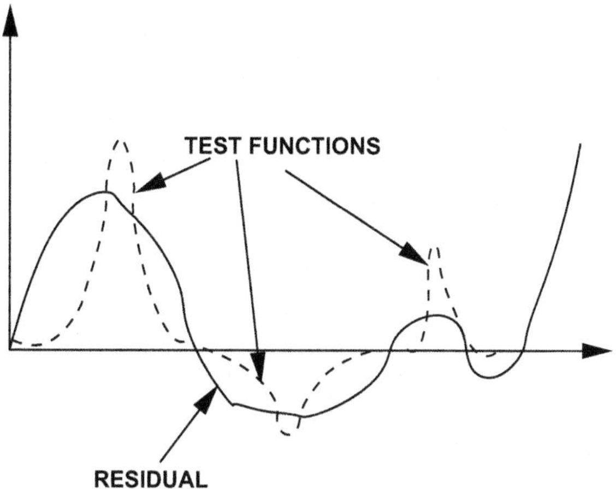

where we choose the ν from an admissible set, to be discussed momentarily. This leads to, $\forall \nu$

$$\int_{\Omega} \frac{d\nu}{dx} \sigma \, dx = \int_{\Omega} f \nu \, dx + \sigma \nu|_{\partial \Omega}, \tag{2.7}$$

On Γ_t, the stress σ on this boundary is known, $\sigma = t^*$ (Fig. 2.1), and is unknown on Γ_u, and thus, we decide to restrict our choices of ν's to those that attain $\nu|_{\Gamma_u} = 0$. We note the use of the symbol t^* stems from the identification of stresses on the boundary as "tractions." Also, choosing a priori for the solution from those functions such that $u|_{\Gamma_u} = u^*$, where u^* is the applied boundary displacement, on a displacement part of the boundary, Γ_u, we have

$$\boxed{\begin{array}{l}\text{Find } u, u|_{\Gamma_u} = u^*, \quad \text{such that } \forall \nu, \nu|_{\Gamma_u} = 0 \\[2ex] \underbrace{\int_{\Omega} \frac{d\nu}{dx} E \frac{du}{dx} \, dx}_{\stackrel{\text{def}}{=} \mathcal{B}(u,\nu)} = \underbrace{\int_{\Omega} f \nu \, dx + t^* \nu|_{\Gamma_t}}_{\stackrel{\text{def}}{=} \mathcal{F}(\nu)}.\end{array}} \tag{2.8}$$

This is called a *weak* form because it does not require the differentiability of σ. In other words, the differentiability requirements have been *weakened*. It is clear that we are able to consider problems with quite irregular solutions. We observe that if we test the solution with all possible test functions of sufficient smoothness, then the weak solution is equivalent to the strong solution. *We emphasize that provided the true solution is smooth enough, the weak and strong forms are equivalent, which can be seen by the above constructive derivation.* To explain the point more clearly, we consider a simple example.

2.3 An Example

Let us define a one-dimensional continuous function $r \in C^0(\Omega)$, on a one-dimensional domain, $\Omega = (0, L)$. Our claim is that

$$\int_{\Omega} r\nu \, dx = 0, \tag{2.9}$$

$\forall \nu \in C^0(\Omega)$, implies $r = 0$. This can be easily proven by contradiction. Suppose $r \neq 0$ at some point $\zeta \in \Omega$. Since $r \in C^0(\Omega)$, there must exist a subdomain (subinterval), $\omega \in \Omega$, defined through δ, $\omega \stackrel{\text{def}}{=} \zeta \pm \delta$ such that r has the same sign as at point ζ. Since ν is arbitrary, we may choose ν to be zero outside of this interval, and with the same sign as r inside (Fig. 2.3). This would imply that

$$0 < \int_{\Omega} r\nu \, dx = \int_{\omega} r\nu \, dx = 0, \tag{2.10}$$

which is a contradiction. Now select

$$r = \frac{d\sigma}{dx} + f \in C^0(\Omega) \Rightarrow \frac{d}{dx}\left(E\frac{du}{dx}\right) + f \in C^0(\Omega) \Rightarrow u \in C^2(\Omega). \tag{2.11}$$

Therefore, for this model problem, the equivalence of weak and strong forms occurs if $u \in C^2(\Omega)$.

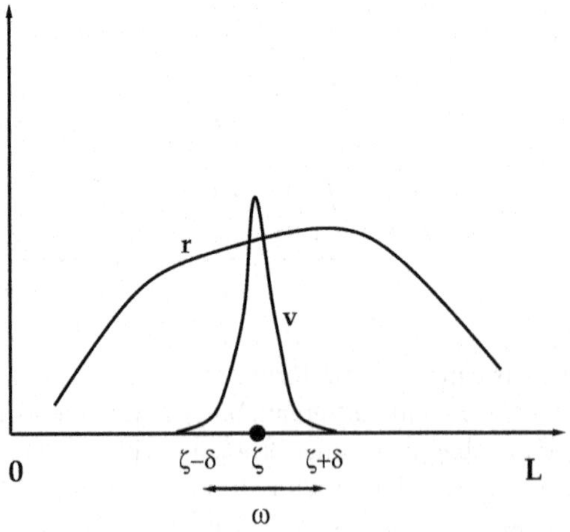

Fig. 2.3 A residual function and a test function

2.4 Some Restrictions

A key question is the selection of the sets of functions in the weak form. Somewhat naively, the answer is simple; the integrals must remain finite. Therefore, the following restrictions hold $(\forall \nu)$, $\int_{\Omega} \frac{d\nu}{dx}\sigma\, dx < \infty$, $\int_{\Omega} f\nu\, dx < \infty$, $\int_{\partial\Omega} t\nu\, dx < \infty$ and govern the selection of the approximation spaces. In order to make precise statements one must have a method of "book keeping." Such a system is to employ so-called Hilbertian Sobolev spaces. We recall that a norm has three main characteristics for any vectors u and ν such that $||u|| < \infty$ and $||\nu|| < \infty$ are (1) $||u|| > 0$, $||u|| = 0$ if and only if $u = 0$ ("positivity"), (2) $||u + \nu|| \leq ||u|| + ||\nu||$ (triangle inequality), and (3) $||\alpha u|| = |\alpha|||u||$, where α is a scalar constant ("scalability"). Certain types of norms, so-called Hilbert space norms, are frequently used in mathematical physics. Following standard notation, we denote $H^1(\Omega)$ as the usual space of scalar functions with generalized partial derivatives of order ≤ 1 in $L^2(\Omega)$; i.e., it is square integrable. In other words, $u \in H^1(\Omega)$ if

$$||u||^2_{H^1(\Omega)} \stackrel{\text{def}}{=} \int_{\Omega} \frac{\partial u}{\partial x}\frac{\partial u}{\partial x}\, dx + \int_{\Omega} uu\, dx < \infty. \tag{2.12}$$

Using these definitions, a complete boundary value problem can be written as follows. The input data loading is assumed to be such that for body forces $f \in L^2(\Omega)$ and boundary traction $\sigma = t^* \in L^2(\Gamma_t)$, but less smooth data can be considered without complications. In summary we assume that our solutions obey these restrictions, leading to the following weak form

$$\boxed{\begin{aligned}&\text{Find } u \in H^1(\Omega), u|_{\Gamma_u} = u^*, \quad \text{such that } \forall \nu \in H^1(\Omega), \nu|_{\Gamma_u} = 0\\ &\int_{\Omega} \frac{d\nu}{dx} E \frac{du}{dx}\, dx = \int_{\Omega} f\nu\, dx + t^*\nu|_{\Gamma_t}.\end{aligned}} \tag{2.13}$$

We note that if the data in (2.13) are smooth and if (2.13) possesses a solution u that is sufficiently regular, then u is the solution of the classical problem in strong form

$$\boxed{\begin{aligned}&\tfrac{d}{dx}(E\tfrac{du}{dx}) + f = 0, \qquad \forall x \in \Omega,\\ &u = u^*, \qquad \forall x \in \Gamma_u,\\ &\sigma = E\tfrac{du}{dx} = t^*, \qquad \forall x \in \Gamma_t.\end{aligned}} \tag{2.14}$$

2.5 Remarks on Nonlinear Problems

The treatment of nonlinear problems is outside the scope of this introductory monograph. However, a few comments are in order. The literature of solving nonlinear problems with the FEM is vast. This is a complex topic that is best illustrated with an extremely simple one-dimensional example with material nonlinearities. Starting with

$$\frac{d}{dx}\underbrace{\left(E(\underbrace{\frac{du}{dx}}_{\stackrel{\text{def}}{=}\epsilon})^p\right)}_{\stackrel{\text{def}}{=}\sigma}+f=0 \tag{2.15}$$

the weak form reads

$$\int_0^L \frac{d\nu}{dx}\sigma\,dx=\int_0^L f\nu\,dx+t^*v|_{\Gamma_t}. \tag{2.16}$$

Using a Taylor series expansion of $\sigma(\epsilon(u))$ about a trial solution $u^{(k)}$ yields (k will be used as an iteration counter)

$$\begin{aligned}\sigma(u^{(k+1)}) &= E(\epsilon(u^{(k+1)}))^p\\ &\approx E\left((\epsilon(u^{(k)}))^p+p(\epsilon(u^{(k)}))^{p-1}\times\left(\epsilon(u^{(k+1)})-\epsilon(u^{(k)})\right)+\mathcal{O}(||u^{(k+1)}-u^{(k)}||^2)\right)\end{aligned} \tag{2.17}$$

and substituting this into the weak form yields

$$\begin{aligned}\int_0^L \frac{d\nu}{dx}\underbrace{\left(Ep(\epsilon(u^{(k)}))^{p-1}\right)}_{E^{tan}}\epsilon(u^{(k+1)})\,dx &= \int_0^L f\nu\,dx+t^*\nu|_{\Gamma_t}\\ &-\int_0^L\frac{d\nu}{dx}E\left((\epsilon(u^{(k)}))^p-p((\epsilon(u^{(k)}))^p)\right)dx.\end{aligned} \tag{2.18}$$

One then iterates $k=1,2,...,$ until $||u^{(k+1)}-u^{(k)}||\leq TOL$. Convergence of such a Newton-type formulation is of concern. We refer the reader to the seminal book of Oden [1], which developed and pioneered nonlinear formulations and convergence analysis. For example, consider a general abstract nonlinear equation of the form

$$\boldsymbol{\Pi}(\boldsymbol{u})=0, \tag{2.19}$$

and the expansion

$$\boldsymbol{\Pi}(\boldsymbol{u}^{(k+1)}) = \boldsymbol{\Pi}(\boldsymbol{u}^{(k)}) + \nabla_{\boldsymbol{u}}\boldsymbol{\Pi}(\boldsymbol{u}^{(k)}) \cdot (\boldsymbol{u}^{(k+1)} - \boldsymbol{u}^{(k)}) + \mathcal{O}(||\boldsymbol{u}^{(k+1)} - \boldsymbol{u}^{(k)}||^2) \approx 0. \tag{2.20}$$

The Newton update can be written in the following form

$$\boldsymbol{u}^{(k+1)} = \boldsymbol{u}^{(k)} - \left(\boldsymbol{\Pi}^{TAN}(\boldsymbol{u}^{(k)})\right)^{-1} \cdot \boldsymbol{\Pi}(\boldsymbol{u}^{(k)}), \tag{2.21}$$

where $\boldsymbol{\Pi}^{TAN}(\boldsymbol{u}) \stackrel{\text{def}}{=} \nabla_{\boldsymbol{u}}\boldsymbol{\Pi}(\boldsymbol{u})$ is the so-called tangent operator. One immediately sees a potential difficulty, due to the possibility of a zero, or near zero, tangent when employing a Newton's method to a system that may have a nonmonotonic response, for example those involving material laws with softening. Specialized techniques can be developed for such problems, and we refer the reader to the state of the art found in Wriggers [2].

References

1. Oden, J. T. (1972). *Finite elements of non-linear continua*. New York: McGraw-Hill.
2. Wriggers, P. (2008). *Nonlinear finite element analysis*. Berlin: Springer.

A Finite Element Implementation in One Dimension

3

3.1 Introduction

Classical techniques construct approximations from globally kinematically admissible functions, which we define as functions that satisfy the displacement boundary condition beforehand. Two main obstacles arise: (1) it may be very difficult to find a kinematically admissible function over the entire domain and (2) if such functions are found they lead to large, strongly coupled and complicated systems of equations. These problems have been overcome by the fact that local approximations (posed over very small partitions of the entire domain) can deliver high-quality solutions and simultaneously lead to systems of equations which have an advantageous mathematical structure amenable to large-scale computation by high-speed computers. These piecewise or "elementwise" approximations have been recognized at least 60 years ago by Courant [1] as being quite advantageous. There have been a variety of such approximation methods to solve equations of mathematical physics. The most popular method of this class is the finite element method (FEM). The central feature of the method is to partition the domain in a systematic manner into an assembly of discrete subdomains or "elements," and then to approximate the solution of each of these pieces in a manner that couples them to form a global solution valid over the whole domain. The process is designed to keep the resulting algebraic systems as computationally manageable, and memory efficient, as possible.

T. I. Zohdi, *A Finite Element Primer for Beginners*, The Basics,
https://doi.org/10.1007/978-3-319-70428-9_3

3.2 Weak Formulation

Consider the following general weak form introduced earlier

$$\boxed{\begin{array}{l}\text{Find } u \in H^1(\Omega)\ u|_{\Gamma_u} = d \text{ such that } \forall \nu \in H^1(\Omega), \nu|_{\Gamma_u} = 0 \\ \displaystyle\int_\Omega \frac{d\nu}{dx} E \frac{du}{dx}\, dx = \int_\Omega f\nu\, dx + t^*\nu|_{\Gamma_t}.\end{array}} \tag{3.1}$$

3.3 FEM Approximation

We approximate u by

$$u^h(x) = \sum_{j=1}^{N} a_j\phi_j(x). \tag{3.2}$$

If we choose ν with the same approximation functions, but a different linear combination

$$\nu^h(x) = \sum_{i=1}^{N} b_i\phi_i(x), \tag{3.3}$$

then we may write

$$\underbrace{\int_\Omega \frac{d}{dx}\left(\sum_{i=1}^{N} b_i\phi_i(x)\right) E \frac{d}{dx}\left(\sum_{j=1}^{N} a_j\phi_j(x)\right) dx}_{\stackrel{\text{def}}{=}\text{stiffness contribution}}$$
$$= \underbrace{\int_\Omega \left(\sum_{i=1}^{N} b_i\phi_i(x)\right) f\, dx}_{\stackrel{\text{def}}{=}\text{body load contribution}} + \underbrace{\left(\left(\sum_{i=1}^{N} b_i\phi_i(x)\right) t^*\right)|_{\Gamma_t}}_{\stackrel{\text{def}}{=}\text{traction load contribution}}. \tag{3.4}$$

Since the ν's are arbitrary, the b_i are arbitrary, i.e., $\forall \nu \Rightarrow \forall b_i$, therefore

$$\boxed{\begin{array}{l}\sum_{i=1}^{N} b_i\left(\sum_{j=1}^{N} K_{ij}a_j - R_i\right) = 0 \Rightarrow [K]\{a\} = \{R\}, \\ K_{ij} \stackrel{\text{def}}{=} \int_\Omega \frac{d\phi_i}{dx} E \frac{d\phi_j}{dx}\, dx \text{ and} \\ R_i \stackrel{\text{def}}{=} \int_\Omega \phi_i f\, dx + \phi_i t^*|_{\Gamma_t},\end{array}} \tag{3.5}$$

where $[K]$ is an $N \times N$ ("stiffness") matrix with components K_{ij} and $\{R\}$ is an $N \times 1$ ("load") vector with components R_i. This is the system of equations that is to be solved. Thus, large N implies large systems of equations and more computational effort. However, with increasing N, we obtain more accurate approximate solutions. We remark that large N does not seem like much of a concern for one-dimensional problems, but is of immense concern for three-dimensional problems.

3.4 Construction of FEM Basis Functions

As mentioned, a primary problem with Galerkin's method is that it provides no systematic way of constructing approximation functions. The difficulties that arise include (1) ill-conditioned systems due to poor choices of approximation functions and (2) domains with irregular geometries. To circumvent these problems, the FEM defines basis (approximation) functions in a piecewise manner over a subdomain, "the finite elements," of the entire domain. The basis functions are usually simple polynomials of low degree. The following three criteria are important:

- The basis functions are smooth enough to be members of $H^1(\Omega)$.
- The basis functions are simple piecewise polynomials, defined element by element.
- The basis functions form a simple nodal basis where $\phi_i(x_j) = 0$ $(i \neq j)$ and $\phi_i(x_i) = 1$, furthermore, $\sum_{i=1}^{N} \phi_i(x) = 1$ for all x and $\phi_i(x) = 0$ outside of the elements that share node i.

A set of candidate functions are defined by

$$\phi(x) = \frac{x - x_{i-1}}{h_i} \qquad \text{for } x_{i-1} \leq x \leq x_i, \tag{3.6}$$

where $h_i = x_i - x_{i-1}$ and

$$\phi(x) = 1 - \frac{x - x_i}{h_{i+1}} \qquad \text{for } x_i \leq x \leq x_{i+1}, \tag{3.7}$$

and $\phi(x) = 0$ otherwise. The derivative of the function is

$$\frac{d\phi}{dx} = \frac{1}{h_i} \qquad \text{for } x_{i-1} \leq x \leq x_i, \tag{3.8}$$

and

$$\frac{d\phi}{dx} = -\frac{1}{h_{i+1}} \qquad \text{for } x_i \leq x \leq x_{i+1}. \tag{3.9}$$

The functions are arranged so that the "apex" of the ith function coincides with the ith node (Fig. 3.1). This framework provides many advantages, for example simple numerical integration.

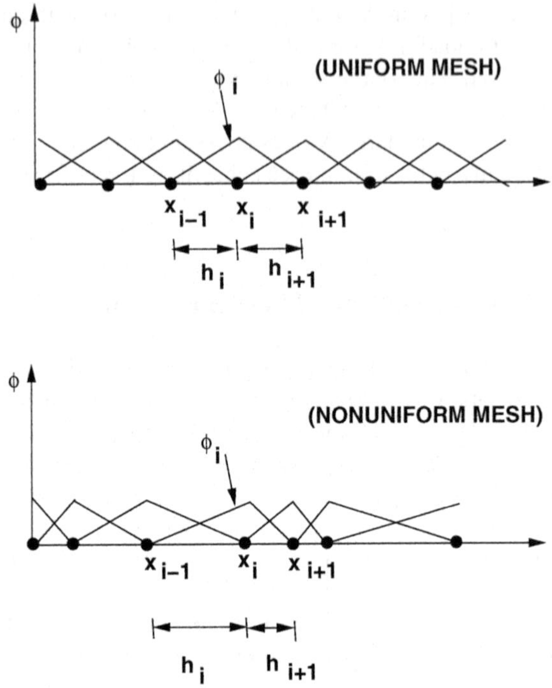

Fig. 3.1 A one-dimensional finite element basis. At the top, is a uniform mesh example and at the bottom, nonuniform

3.5 Integration and Gaussian Quadrature

Gauss made the crucial observation that one can integrate a polynomial of order 2G-1 exactly with G "sampling" points and appropriate weights. Thus, in order to automate the integration process, one redefines the function $F(x)$ over a normalized unit domain $-1 \leq \zeta \leq +1$

$$\int_0^L F(x)\,dx = \int_{-1}^{1} F(x(\zeta))\,J(\zeta)d\zeta = \sum_{i=1}^{G} w_i F(\zeta_i)J(\zeta_i) = \sum_{i=1}^{G} w_i \hat{F}(\zeta_i), \tag{3.10}$$

where J is the Jacobian of the transformation. Unlike most integration schemes, Gaussian quadrature relaxes the usual restriction that the function sampling locations be evenly spaced. According to the above, we should be able to integrate a cubic (and lower order) term exactly with $G = 2$ points, since $(2G - 1) = 3$. Therefore

- For a cubic (ζ^3):

$$\int_{-1}^{1} \zeta^3\,d\zeta = 0 = w_1\zeta_1^3 + w_2\zeta_2^3 \tag{3.11}$$

- For a quadratic (ζ^2):

$$\int_{-1}^{1} \zeta^2 \, d\zeta = 2/3 = w_1\zeta_1^2 + w_2\zeta_2^2 \tag{3.12}$$

- For a linear (ζ):

$$\int_{-1}^{1} \zeta \, d\zeta = 0 = w_1\zeta_1 + w_2\zeta_2 \tag{3.13}$$

- For a constant (1):

$$\int_{-1}^{1} 1 \, d\zeta = 2 = w_1 1 + w_2 1 = w_1 + w_2 \tag{3.14}$$

There are four variables, $\zeta_1, \zeta_2, w_1, w_2$, to solve for. The solution that satisfies all of the requirements is $\zeta_1 = \sqrt{1/3} = -\zeta_2$ and $w_1 = w_2 = 1$. For the general case of G points, we have

$$\int_{-1}^{1} \hat{F}(\zeta) d\zeta = \sum_{i=1}^{G} w_i \hat{F}(\zeta_i) \tag{3.15}$$

and subsequently 2G nonlinear equations for the ζ_i's and w_i's. Fortunately, the ζ_i's are the roots to the Gth degree Legendre polynomial, defined via the recursion (Fig. 3.2)

$$(G+1)L_{G+1}(\zeta) - (2G+1)\zeta L_G(\zeta) + GL_{G-1}(\zeta) = 0, \tag{3.16}$$

with $L_o(\zeta) = 1$, $L_1(\zeta) = \zeta$. The roots of the Legendre polynomial are well known and tabulated. Once the roots are determined the remaining equations for the w_i's are linear and easy to solve. Fortunately, the roots are precomputed over a normalized unit domain, and one does not need to compute them. The only task is to convert the domain of each element to a standard unit domain (in the next section). A table of Gauss weights can be found in Table 3.1.

3.5.1 An Example

Consider the following integral

$$I \stackrel{\text{def}}{=} \int_{0.2}^{1.5} 10e^{-x^2} \, dx. \tag{3.17}$$

This integral is of the form

$$I \stackrel{\text{def}}{=} \int_a^b f(x) \, dx = \int_{-1}^{1} f\left(\frac{(b-a)\zeta + b + a}{2}\right) \underbrace{\frac{(b-a)}{2}}_{J} \, d\zeta, \tag{3.18}$$

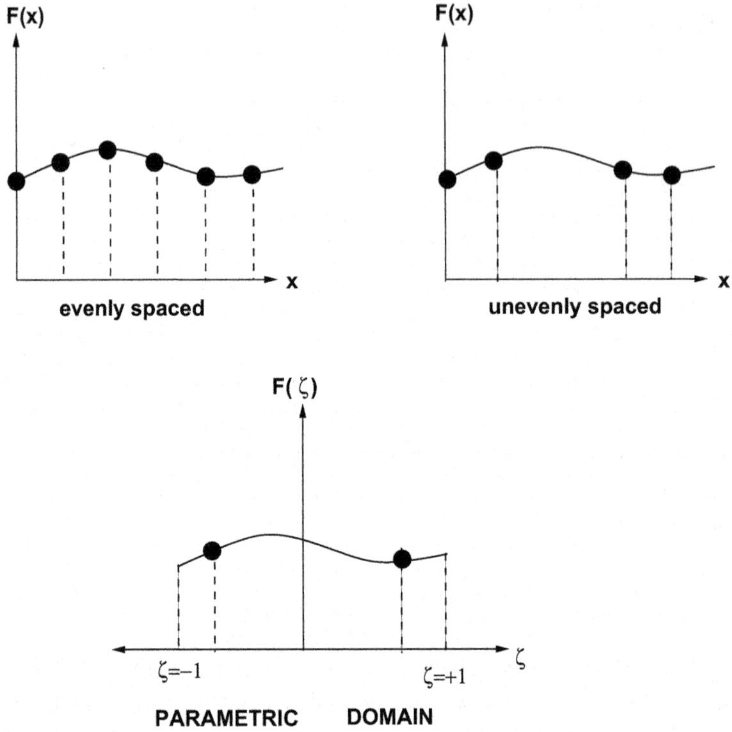

Fig. 3.2 Integration using Gaussian quadrature

Table 3.1 Gauss integration rules

Gauss rule	ζ_i	w_i
2	0.577350269189626	1.000000000000000
	−0.577350269189626	1.000000000000000
3	0.000000000000000	0.888888888888889
	0.774596669224148	0.555555555555556
	−0.774596669224148	0.555555555555556
4	0.339981043584856	0.652145154862546
	0.861136311594053	0.347854845137454
	−0.339981043584856	0.652145154862546
	−0.861136311594053	0.347854845137454
5	0.000000000000000	0.568888888888889
	0.538469310105683	0.478628670499366
	0.906179845938664	0.236926885056189
	−0.538469310105683	0.478628670499366
	−0.906179845938664	0.236926885056189

where we have the following mapping

$$x = \frac{(b-a)\zeta + b + a}{2} \Rightarrow dx = \frac{b-a}{2} d\zeta. \tag{3.19}$$

Applying this transformation, we have

$$I \stackrel{\text{def}}{=} \int_{0.2}^{1.5} 10e^{-x^2}\, dx = \frac{1.5-0.2}{2} \int_{-1}^{1} 10e^{-(0.65\zeta+0.85)^2}\, d\zeta, \tag{3.20}$$

where $x = 0.65\zeta + 0.85$. Applying a three-point rule yields (*the exact answer is* 6.588)

$$\begin{aligned} I &= \frac{1.5-0.2}{2} \int_{-1}^{1} 10e^{-(0.65\zeta+0.85)^2}\, d\zeta \\ &= 6.5 \left(0.5555e^{-(0.65(-0.77459)+0.85)^2} + 0.8888e^{-(0.65(0)+0.85)^2} + 0.5555e^{-(0.65(0.77459)+0.85)^2}\right) \\ &= 6.586. \end{aligned} \tag{3.21}$$

3.6 Global/Local Transformations

One strength of the finite element method is that most of the computations can be done in an element-by-element manner. Accordingly, we define the entries of the stiffness matrix $[K]$ as

$$K_{ij} = \int_{\Omega} \frac{d\phi_i}{dx} E \frac{d\phi_j}{dx}\, dx, \tag{3.22}$$

and the load vector as

$$R_i = \int_{\Omega} \phi_i f\, dx + \phi_i t^*|_{\Gamma_t}. \tag{3.23}$$

We partition the domain Ω into elements, $\Omega_1, \Omega_2, ..., \Omega_e, ...\Omega_N$, and can consequently break the calculations (integrals over Ω) into elements (integrals over Ω_e), $K_{ij} = \sum_e K_{ij}^e$, where

$$K_{ij}^e = \int_{\Omega_e} \frac{d\phi_i}{dx} E \frac{d\phi_j}{dx}\, dx \tag{3.24}$$

and

$$R_i^e = \int_{\Omega_e} \phi_i f\, dx + \phi_i t^*|_{\Gamma_{t,e}}, \tag{3.25}$$

where $R_i = \sum_e R_i^e$ and $\Gamma_{t,e} = \Gamma_t \cap \Omega_e$.

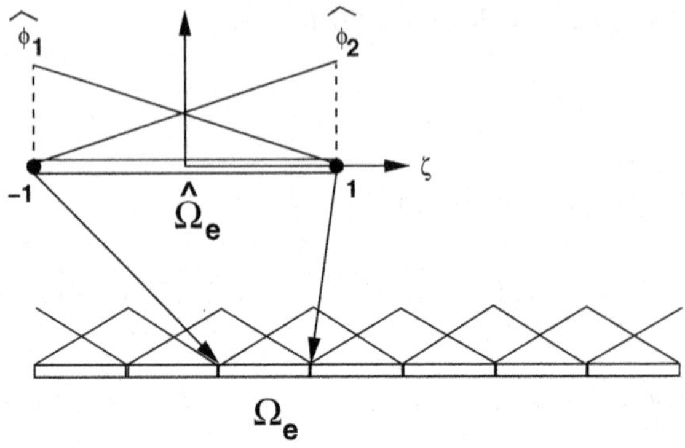

Fig. 3.3 A one-dimensional linear finite element mapping

In order to make the calculations systematic we wish to use the generic or master element defined in a local coordinate system (ζ). Accordingly, we need the following mapping functions, from the master coordinates to the real spatial coordinates, $M_x(\zeta) \mapsto x$ (Fig. 3.3)

$$x = \sum_{i=1}^{2} \mathcal{X}_i \hat{\phi}_i \stackrel{\text{def}}{=} M_x(\zeta), \tag{3.26}$$

where the $\mathcal{X}_i$ are the true spatial coordinates of the ith node, and where $\hat{\phi}(\zeta) \stackrel{\text{def}}{=} \phi(x(\zeta))$. These types of mappings are usually termed "parametric" maps. If the polynomial order of the shape functions is as high as the Galerkin approximation over the element, it is called an "isoparametric" map, lower, then "subparametric" map, higher, then "superparametric".

3.7 Differential Properties of Shape Functions

The master element shape functions form a nodal bases of linear approximation given by

$$\hat{\phi}_1 = \frac{1}{2}(1 - \zeta) \quad \text{and} \quad \hat{\phi}_2 = \frac{1}{2}(1 + \zeta). \tag{3.27}$$

They have the following properties:

- For linear elements we have a nodal basis consisting of two nodes, and thus two degrees of freedom.
- The nodal shape functions can be derived quite easily, by realizing that it is a nodal basis; i.e., they are unity at the corresponding node and zero at all other nodes.

We note that the ϕ_i's are never really computed; we actually start with the $\hat{\phi}_i$'s and then map them into the actual problem domain. Therefore in the stiffness matrix and right-hand side element calculations, all terms must be defined in terms of the local coordinates. With this in mind, we introduce some fundamental quantities, such as the finite element mapping deformation gradient

$$F \stackrel{\text{def}}{=} \frac{dx}{d\zeta}. \tag{3.28}$$

The corresponding one-dimensional determinant is $|F| = \frac{dx}{d\zeta} \stackrel{\text{def}}{=} J$, which is known as the Jacobian. We will use $|F|$ and J interchangeably throughout this monograph. The differential relations $\zeta \to x$ are

$$\frac{d()}{d\zeta} = \frac{dx}{d\zeta}\frac{d()}{dx} = J\frac{d()}{dx}. \tag{3.29}$$

The inverse differential relations $x \to \zeta$ are

$$\frac{d()}{dx} = \frac{d\zeta}{dx}\frac{d()}{d\zeta} = \frac{1}{J}\frac{d()}{d\zeta}. \tag{3.30}$$

We can now express $\frac{d}{dx}$ in terms ζ, via

$$\frac{d\phi}{dx} = \frac{d}{dx}\phi(M(\zeta)) = \frac{d\zeta}{dx}\frac{d}{d\zeta}\phi(M(\zeta)) = \frac{d\zeta}{dx}\frac{d}{d\zeta}\hat{\phi}(\zeta). \tag{3.31}$$

Finally with quadrature for each element

$$K^e_{ij} = \sum_{q=1}^{g} \underbrace{w_q \left(\frac{d}{d\zeta}(\phi_i(M(\zeta))\right)\frac{d\zeta}{dx} E \left(\frac{d}{d\zeta}(\phi_j(M(\zeta))\right)\frac{d\zeta}{dx}|F|}_{\text{evaluated at } \zeta=\zeta_q} \tag{3.32}$$

and

$$R^e_i = \sum_{q=1}^{g} \underbrace{w_q\phi_i(M(\zeta))f|F|}_{\text{evaluated at } \zeta=\zeta_q} + \underbrace{\phi_i(M(\zeta))t^*}_{\text{evaluated on traction endpoints}}, \tag{3.33}$$

where the w_q are Gauss weights.

Remarks: It is permitted to have material discontinuities within the finite elements. On the implementation level, the system of equations to be solved is $[K]\{a\} = \{R\}$, where the stiffness matrix is represented by $K(I, J)$, where (I, J) are the global entries. However, one can easily take advantage of the element-by-element structure and store the entries via $k^e(e, i, j)$, where (e, i, j) are the local (element) entries. For the local storage approach, a global/local index relation must be made to connect the local entry to the global entry when the linear algebraic solution process begins. This is a relatively simple and efficient storage system to encode. The element-by-element strategy has other advantages with regard to element-by-element system solvers. This is trivial in one dimension; however, it can be complicated in three dimensions. This is discussed later.

3.8 Post-Processing

Post-processing for the stress, strain, and energy from the existing displacement solution, i.e., the values of the nodal displacements, the shape functions, are straightforward. Essentially the process is the same as the formation of the weak form in the system. Therefore, for each element

$$\frac{du}{dx} = \frac{d}{dx}\sum_{i=1}^{2} a_i \phi_i = \left(\frac{d}{d\zeta}\sum_{i=1}^{2} a_i \hat{\phi}_i\right)\frac{d\zeta}{dx}. \tag{3.34}$$

3.9 A Detailed Example

3.9.1 Weak Form

Consider the following problem (Fig. 3.4)

$$\frac{d}{dx}(E(x)\frac{du}{dx}) + f(x) = 0, \tag{3.35}$$

$u(0) = 0$ and $\frac{du}{dx}(1) = t$, posed over a domain of unit length. The weak form is

$$\int_o^{L=1} \frac{dv}{dx} E(x) \frac{du}{dx}\, dx = \int_o^{L=1} f(x) v\, dx + \underbrace{(E(x)\frac{du}{dx} v)|_0^1}_{=t^* v}. \tag{3.36}$$

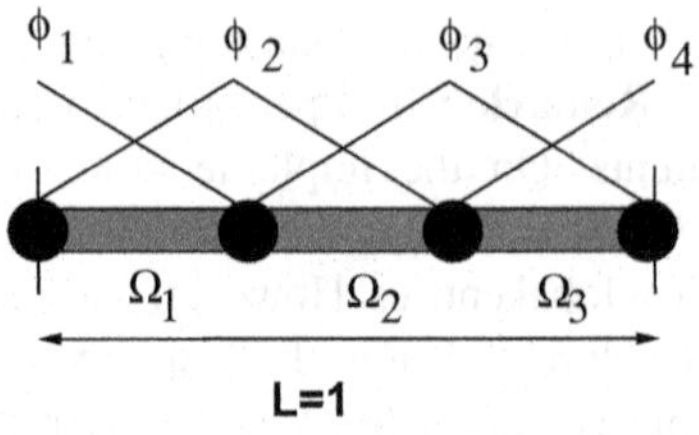

Fig. 3.4 Three elements and four nodes

Using three elements (four nodes), each of equal size, the following holds:

- Over element 1 (Ω_1): $\mathcal{X}_i = \mathcal{X}_1 = 0$ and $\mathcal{X}_{i+1} = \mathcal{X}_2 = 1/3$, $\phi_1(x) = 1 - 3x$ and $\phi_2(x) = 3x$,
- Over element 2 (Ω_2): $\mathcal{X}_i = \mathcal{X}_2 = 1/3$ and $\mathcal{X}_{i+1} = \mathcal{X}_3 = 2/3$, $\phi_2(x) = 2 - 3x$ and $\phi_3(x) = -1 + 3x$,
- Over element 3 (Ω_3): $\mathcal{X}_i = \mathcal{X}_3 = 2/3$ and $\mathcal{X}_{i+1} = \mathcal{X}_4 = 1$, $\phi_3(x) = 3 - 3x$ and $\phi_4(x) = -2 + 3x$,

We break the calculations up element by element. *All calculations between* $0 \leq x \leq 1/3$ *belong to element number 1, while all calculations between* $1/3 \leq x \leq 2/3$ *belong to element number 2 and all calculations between* $2/3 \leq x \leq 1$ *belong to element number 3.*

3.9.2 Formation of the Discrete System

For element number 1, to compute $K_{ij}^{e=1}$, we study the following term for $i = 1, 2, 3$:

$$\sum_{j=1}^{N} \left(\int_0^{1/3} \frac{d\phi_i}{dx} E(x) \frac{d\phi_j}{dx} \, dx \right) a_j. \tag{3.37}$$

Explicitly, for $i = 1$, we have

$$\underbrace{\left(\int_0^{1/3} \frac{d\phi_1}{dx} E(x) \frac{d\phi_1}{dx} \, dx \right)}_{K_{11}^{e=1}} a_1 + \underbrace{\left(\int_0^{1/3} \frac{d\phi_1}{dx} E(x) \frac{d\phi_2}{dx} \, dx \right)}_{K_{12}^{e=1}} a_2 + \underbrace{\left(\int_0^{1/3} \frac{d\phi_1}{dx} E(x) \frac{d\phi_3}{dx} \, dx \right)}_{=0} a_3 + 0, \; etc., \tag{3.38}$$

where the zero-valued terms vanish because the basis functions are zero over the first finite element domain. The entries such as $K_{ij}^{e=1}$ multiply the term a_j, which dictate their location within the global stiffness matrix. If we repeat the procedure for $i = 2$, $j = 1, 2, 3$, we obtain the entries for the global stiffness matrix (4×4)

$$\begin{bmatrix} K_{11}^{e=1} & K_{12}^{e=1} & 0 & 0 \\ K_{21}^{e=1} & K_{22}^{e=1} & 0 & 0 \\ 0 & 0 & 0 & 0 \\ 0 & 0 & 0 & 0 \end{bmatrix} \tag{3.39}$$

stemming from the placement of the local element stiffness matrix

$$\begin{bmatrix} K_{11}^{e=1} & K_{12}^{e=1} \\ K_{21}^{e=1} & K_{22}^{e=1} \end{bmatrix} \tag{3.40}$$

into the global stiffness matrix. Following a similar procedure for the right-hand side (load)

$$\underbrace{\left(\int_0^{1/3} \phi_i f(x)\, dx\right)}_{R_i^{e=1}} \tag{3.41}$$

yields ($i = 1, 2$)

$$\begin{bmatrix} R_1^{e=1} \\ R_2^{e=1} \end{bmatrix}. \tag{3.42}$$

Repeating the procedure for all three of the elements yields

$$\begin{bmatrix} K_{11}^{e=1} & K_{12}^{e=1} & 0 & 0 \\ K_{21}^{e=1} & K_{22}^{e=1} + K_{11}^{e=2} & K_{12}^{e=2} & 0 \\ 0 & K_{21}^{e=2} & K_{22}^{e=2} + K_{11}^{e=3} & K_{12}^{e=3} \\ 0 & 0 & K_{21}^{e=3} & K_{22}^{e=3} \end{bmatrix} \tag{3.43}$$

and

$$\begin{bmatrix} R_1^{e=1} \\ R_2^{e=1} + R_1^{e=2} \\ R_2^{e=2} + R_1^{e=3} \\ R_2^{e=3} \end{bmatrix}. \tag{3.44}$$

Note that the load vector

$$R_2^{e=3} = \int_{2/3}^{1} \phi_4 f(x)\, dx + E(x)\frac{du}{dx}\phi_4(1) = \int_{2/3}^{1} \phi_4 f(x)\, dx + t^* \tag{3.45}$$

has a traction contribution from the right endpoint. In summary, the basic process is to (1) compute element by element and (2) to sweep over all basis function contributions over each element.

Remark: We note that all integrals are computed using Gaussian quadrature.

3.9.3 Applying Boundary Conditions

Applying the primal (displacement) boundary conditions requires us to recall that the b_i's in the representation of the test functions are not arbitrary at the endpoints, thus the equations associated with those test functions have to be eliminated, and the value of the approximate solution enforced at the displacement boundary condition via[1]

$$u^h(x=0) = \sum_{j=1}^{4} a_j \phi_j(x=0) = a_1, \tag{3.46}$$

which is the displacement-specified boundary condition. Thus, we have the following system of equations

$$\begin{bmatrix} K_{22}^{e=1} + K_{11}^{e=2} & K_{12}^{e=2} & 0 \\ K_{21}^{e=2} & K_{22}^{e=2} + K_{11}^{e=3} & K_{12}^{e=3} \\ 0 & K_{21}^{e=3} & K_{22}^{e=3} \end{bmatrix} \begin{bmatrix} a_2 \\ a_3 \\ a_4 \end{bmatrix} = \begin{bmatrix} R_2^{e=1} + R_1^{e=2} - K_{12}^{e=1} a_1 \\ R_2^{e=2} + R_1^{e=3} \\ R_2^{e=3} \end{bmatrix} \tag{3.47}$$

3.9.4 Massive Data Storage Reduction

The direct storage of $K(I, J)$ requires $N \times N$ entries. The element-by-element storage, $k^e(e, i, j)$, requires $4e$. The memory requirements for an element-by-element paradigm are much smaller than those for a direct scheme, which store needless zeros. For example, for $N = 10^4$ nodes, the direct storage is $(10^4)^2 = 10^8$, while the element-by-element storage is 9999×4, *which is essentially 2500 times less than direct storage.* Additionally, there is a massive reduction of mathematical operations during the algebraic solution phase, because of the element-by-element structure of FEM system.

3.10 Quadratic Elements

In many cases, if the character of the exact solution is known to be smooth, it is advantageous to use higher-order approximation elements. Generally, if the exact solution to a problem is smooth, for sufficiently fine meshes, if one compares, for the same number of nodes, the solution produced with linear basis functions to the solution produced with quadratic basis functions, the quadratically produced solution is more accurate. Similarly, if the exact solution is rough (nonsmooth), for sufficiently fine meshes, if one compares, for the same number of nodes, the solution produced with linear basis functions to the solution produced with quadratic basis functions, the linearly produced solution is more accurate (Fig. 3.5).

To illustrate how to construct a quadratic finite element approximation, we follow a similar template for linear elements, however, with three nodes instead of two. Consistent with the basic nodal basis construction, the basis function must equal

[1]The traction boundary conditions are automatically accounted for in the weak formulation.

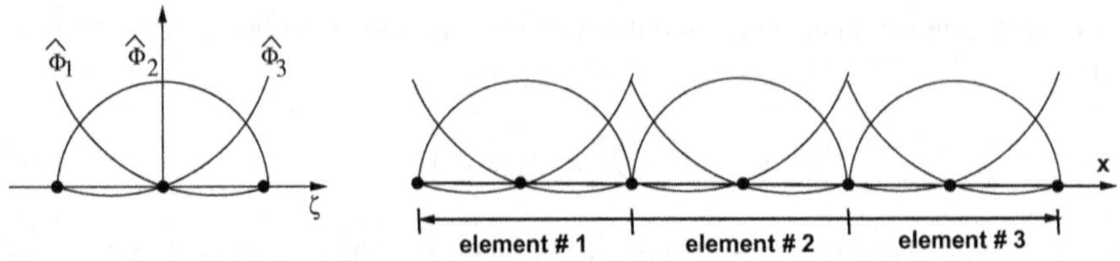

Fig. 3.5 Three quadratic elements with seven nodes

unity on the node it belongs and be zero at the others. Thus, for a generic quadratic element:

- For node # 1: $\hat{\phi}_1(\zeta) = -\frac{1}{2}(1-\zeta)\zeta$, which yields $\hat{\phi}_1(-1) = 1$, $\hat{\phi}(0) = 0$, $\hat{\phi}_1(1) = 0$,
- For node # 2: $\hat{\phi}_2(\zeta) = (1+\zeta)(1-\zeta)$, which yields $\hat{\phi}_2(-1) = 0$, $\hat{\phi}_2(0) = 1$, $\hat{\phi}_2(1) = 0$ and
- For node # 3: $\hat{\phi}_3(\zeta) = \frac{1}{2}(\zeta+1)\zeta$ which yields $\hat{\phi}_3(-1) = 0, \hat{\phi}_3(0) = 0, \hat{\phi}_3(1) = 1$.

Following the approach for linear elements, the connection between x and ζ is

$$x(\zeta) = \mathcal{X}_i\hat{\phi}_1(\zeta) + \mathcal{X}_{i+1}\hat{\phi}_2(\zeta) + \mathcal{X}_{i+2}\hat{\phi}_3(\zeta). \tag{3.48}$$

Clearly, the weak form does not change for linear or quadratic approximations. Furthermore, the quadratically generated system has a similar form to the linearly generated system

$$\sum_{j=1}^{N} K_{ij}a_j = R_i \qquad i = 1, 2, ...N, \tag{3.49}$$

where N is the number of nodes in 1-D. Let us consider an example with three elements, resulting in 7 nodes. Breaking up the integral into the elements

$$\int_0^1 = \int_0^{1/3} + \int_{1/3}^{2/3} + \int_{2/3}^{1}. \tag{3.50}$$

For element #1, for $i = 1, 2...N$, we need to compute

$$\sum_{j=1}^{N} \underbrace{\int_0^{1/3} \frac{d\phi_i}{dx} E(x) \frac{d\phi_j}{dx}\, dx}_{K_{ij}^{e=1}}, \tag{3.51}$$

yielding

$$\sum_{j=1}^{N}\underbrace{\int_0^{1/3}\frac{d\phi_1}{dx}E(x)\frac{d\phi_j}{dx}\,dx}_{K_{1j}^{e=1}}=\underbrace{\int_0^{1/3}\frac{d\phi_1}{dx}E(x)\frac{d\phi_1}{dx}\,dx}_{K_{11}^{e=1}}+\underbrace{\int_0^{1/3}\frac{d\phi_1}{dx}E(x)\frac{d\phi_2}{dx}\,dx}_{K_{12}^{e=1}}+\underbrace{\int_0^{1/3}\frac{d\phi_1}{dx}E(x)\frac{d\phi_3}{dx}\,dx}_{K_{13}^{e=1}}+\underbrace{\int_0^{1/3}\frac{d\phi_1}{dx}E(x)\frac{d\phi_4}{dx}\,dx}_{K_{14}^{e=1}=0}. \tag{3.52}$$

For the right-hand side, for $i = 1, 2...N$, we need to compute

$$\int_0^{1/3}\phi_i f(x)\,dx = R_i^{e=1}, \tag{3.53}$$

thus

$$R_1^{e=1}=\int_0^{1/3}\phi_1 f(x)\,dx. \tag{3.54}$$

Repeating this for $i = 2, 3...N$, we have

$$\begin{bmatrix} K_{11}^{e=1} & K_{12}^{e=1} & K_{13}^{e=1} & 0\,0\,0\,0 \\ K_{21}^{e=1} & K_{22}^{e=1} & K_{23}^{e=1} & 0\,0\,0\,0 \\ K_{31}^{e=1} & K_{32}^{e=1} & K_{33}^{e=1} & 0\,0\,0\,0 \\ 0 & 0 & 0 & 0\,0\,0\,0 \\ 0 & 0 & 0 & 0\,0\,0\,0 \\ 0 & 0 & 0 & 0\,0\,0\,0 \\ 0 & 0 & 0 & 0\,0\,0\,0 \end{bmatrix}\begin{bmatrix} a_1 \\ a_2 \\ a_3 \\ a_4 \\ a_5 \\ a_6 \\ a_7 \end{bmatrix}=\begin{bmatrix} R_1^{e=1} \\ R_2^{e=1} \\ R_3^{e=1} \\ 0 \\ 0 \\ 0 \\ 0 \end{bmatrix} \tag{3.55}$$

This is then repeated for elements 2 and 3, to yield

$$\begin{bmatrix} K_{11}^{e=1} & K_{12}^{e=1} & K_{13}^{e=1} & 0 & 0 & 0 & 0 \\ K_{21}^{e=1} & K_{22}^{e=1} & K_{23}^{e=1} & 0 & 0 & 0 & 0 \\ K_{31}^{e=1} & K_{32}^{e=1} & K_{33}^{e=1}+K_{11}^{e=2} & K_{12}^{e=2} & K_{13}^{e=2} & 0 & 0 \\ 0 & 0 & K_{21}^{e=2} & K_{22}^{e=2} & K_{23}^{e=2} & 0 & 0 \\ 0 & 0 & K_{31}^{e=2} & K_{32}^{e=2} & K_{33}^{e=2}+K_{11}^{e=3} & K_{12}^{e=3} & K_{13}^{e=3} \\ 0 & 0 & 0 & 0 & K_{21}^{e=3} & K_{22}^{e=3} & K_{23}^{e=3} \\ 0 & 0 & 0 & 0 & K_{31}^{e=3} & K_{32}^{e=3} & K_{33}^{e=3} \end{bmatrix}\begin{bmatrix} a_1 \\ a_2 \\ a_3 \\ a_4 \\ a_5 \\ a_6 \\ a_7 \end{bmatrix}=\begin{bmatrix} R_1^{e=1} \\ R_2^{e=1} \\ R_3^{e=1}+R_1^{e=2} \\ R_2^{e=2} \\ R_3^{e=2}+R_1^{e=3} \\ R_2^{e=3} \\ R_3^{e=3} \end{bmatrix} \tag{3.56}$$

One then applies boundary conditions in the same manner as for linear elements.

Remark: A logical question to ask is what is the accuracy of the finite element method? This is addressed in the next chapter.

Reference

1. Courant, R. (1943). Variational methods for the solution of problems of equilibrium and vibrations. *Bulletin of the American Mathematical Society*, *49*, 1–23.

Accuracy of the Finite Element Method in One Dimension

4

4.1 Introduction

As we have seen, the essential idea in the finite element method is to select a finite dimensional subspatial approximation of the true solution and to form the following weak boundary problem

$$\boxed{\begin{aligned}&\text{Find } u^h \in H_u^h(\Omega) \subset H^1(\Omega), \text{ with } u^h|_{\Gamma_u} = d, \text{ such that}\\ &\underbrace{\int_\Omega \frac{d\nu^h}{dx} E \frac{du^h}{dx}\, dx}_{\mathcal{B}(u^h,\nu^h)} = \underbrace{\int_\Omega f\nu^h\, dx + \nu^h t^*|_{\Gamma_t}}_{\mathcal{F}(\nu^h)},\\ &\forall \nu^h \in H_\nu^h(\Omega) \subset H^1(\Omega), \text{ with } \nu^h|_{\Gamma_u} = 0,\end{aligned}} \tag{4.1}$$

where we refer to $H_u^h(\Omega)$ and $H_\nu^h(\Omega)$ as the space of approximations (e.g., linear functions). The critical point is that $H_u^h(\Omega), H_\nu^h(\Omega) \subset H^1(\Omega)$. This "inner" approximation allows the development of straightforward subspatial error estimates. We will choose $H_u^h(\Omega)$ and $H_\nu^h(\Omega)$ to coincide. We have, for any $H^1(\Omega)$ admissible function w, a definition of the so-called energy semi-norm

$$||u - w||_{E(\Omega)}^2 \stackrel{\text{def}}{=} \int_\Omega (\frac{du}{dx} - \frac{dw}{dx}) E (\frac{du}{dx} - \frac{dw}{dx})\, dx = \mathcal{B}(u - w, u - w). \tag{4.2}$$

Note that in the event that nonuniform displacements are specified on the boundary (no rigid motion produced), then $u - w = constant$ is unobtainable unless $u - w = 0$, and the semi-norm in Eq. (4.2) is a norm in the strict mathematical sense. Under relatively mild assumptions, a fundamental a priori error estimate for the finite element method is

$$||u - u^h||_{E(\Omega)} \leq C(u, p) h^{min(r-1,p)} \stackrel{\text{def}}{=} \gamma, \tag{4.3}$$

T. I. Zohdi, *A Finite Element Primer for Beginners*, The Basics,
https://doi.org/10.1007/978-3-319-70428-9_4

where p is the (complete) polynomial order of the finite element method used, r is the regularity of the exact solution, and C is a constant dependent on the exact solution and the polynomial approximation. C is independent of h, the maximum element size in the mesh. For details, see, for example, Ainsworth and Oden [1], Becker, Carey and Oden [2], Carey and Oden [3], Oden and Carey [4], Hughes [5], Szabo and Babuska [6], and Bathe [7].

Remark 1: We note that set of functions specified by $H_u^h(\Omega) \subset H^1(\Omega)$ with $u^h|_{\Gamma_u} = d$ is technically not a space of functions and should be characterized as "a linear variety." This does not pose a problem for the ensuing analysis. For precise mathematical details, see Oden and Demkowicz [8].

Remark 2: We note that $\sqrt{B(u,u)}$ is a norm since:

- Positivity:

$$||u||^2_{E(\Omega)} = B(u,u) \geq 0 \tag{4.4}$$

where, provided that $u \neq constant$, $B(u,u) = 0$ if and only if $u = 0$.

- Triangle inequality:

$$\begin{aligned} ||u+v||^2_{E(\Omega)} &= B(u+v, u+v) \\ &= B(u,u) + 2B(u,v) + B(v,v) \\ &\leq ||u||^2_{E(\Omega)} + 2||u||_{E(\Omega)}||v||_{E(\Omega)} + ||v||^2_{E(\Omega)} \\ &= (||u||_{E(\Omega)} + ||v||_{E(\Omega)})^2. \end{aligned} \tag{4.5}$$

- Scalability by a scalar constant multiplier:

$$||\alpha u||_{E(\Omega)} = \sqrt{B(\alpha u, \alpha u)} = |\alpha|\sqrt{B(u,u)}. \tag{4.6}$$

4.2 The "Best Approximation" Theorem

The FEM solution is optimal in the energy norm. To prove this we use

$$\mathcal{B}(u,\nu) = \mathcal{F}(\nu), \tag{4.7}$$

$\forall \nu \in H^1(\Omega)$ and

$$\mathcal{B}(u^h,\nu^h) = \mathcal{F}(\nu^h), \tag{4.8}$$

$\forall \nu^h \in H_\nu^h(\Omega) \subset H^1(\Omega)$. Subtracting Eq. 4.8 from 4.7 implies a Galerkin-like (Fig. 1.1) orthogonality property of "inner approximations":

$$\mathcal{B}(u-u^h,\nu^h) = \mathcal{B}(e^h,\nu^h) = 0, \qquad \forall \nu^h \in H_\nu^h(\Omega) \subset H^1(\Omega), \tag{4.9}$$

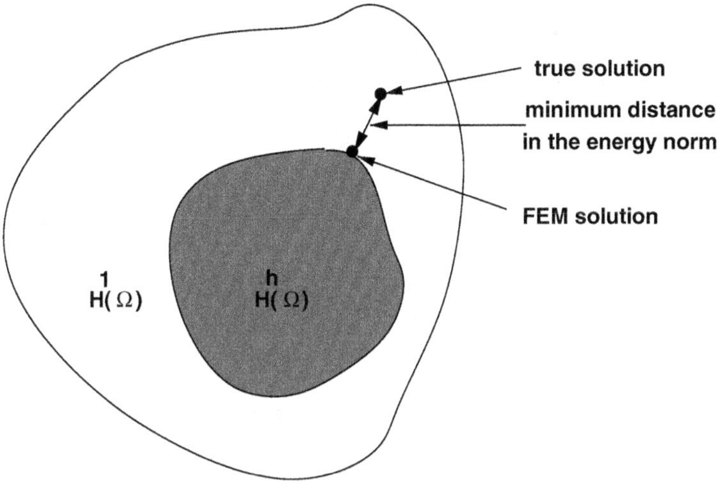

Fig. 4.1 A schematic of the best approximation theorem

where the error is defined by $e^h \stackrel{\text{def}}{=} u - u^h$. An important observation is that any member of the subspace can be represented by

$$e^h - \nu^h = u - u^h - \nu^h = u - z^h, \tag{4.10}$$

where z^h is kinematically admissible. Using this representation we have

$$\mathcal{B}(e^h - \nu^h, e^h - \nu^h) = \mathcal{B}(e^h, e^h) - 2\mathcal{B}(e^h, \nu^h) + \mathcal{B}(\nu^h, \nu^h), \tag{4.11}$$

which implies

$$\mathcal{B}(u - u^h, u - u^h) \leq \mathcal{B}(u - z^h, u - z^h). \tag{4.12}$$

This is called the best approximation theorem (see Fig. 4.1 for a schematic).

4.3 The Principle of Minimum Potential Energy

A useful set of concepts in mathematical physics are minimum principles. By direct manipulation we have ($w \subset H^1(\Omega)$ is any kinematically admissible function)

$$\begin{aligned}||u - w||^2_{E(\Omega)} &= \mathcal{B}(u - w, u - w)\\ &= \mathcal{B}(u, u) + \mathcal{B}(w, w) - 2\mathcal{B}(u, w)\\ &= \mathcal{B}(w, w) - \mathcal{B}(u, u) - 2\mathcal{B}(u, w) + 2\mathcal{B}(u, u)\end{aligned}$$

$$
\begin{aligned}
&= \mathcal{B}(w, w) - \mathcal{B}(u, u) - 2\mathcal{B}(u, w - u)\\
&= \mathcal{B}(w, w) - \mathcal{B}(u, u) - 2\mathcal{F}(w - u)\\
&= \mathcal{B}(w, w) - 2\mathcal{F}(w) - (\mathcal{B}(u, u) - 2\mathcal{F}(u))\\
&= 2\mathcal{J}(w) - 2\mathcal{J}(u),
\end{aligned} \tag{4.13}
$$

where we define the "potential" via

$$
\mathcal{J}(w) \stackrel{\text{def}}{=} \frac{1}{2}\mathcal{B}(w, w) - \mathcal{F}(w) = \frac{1}{2}\int_{\Omega} \frac{dw}{dx} E \frac{dw}{dx}\, dx - \int_{\Omega} f w\, dx - t^{*} w|_{\Gamma_t}. \tag{4.14}
$$

This implies

$$
0 \leq ||u - w||^2_{E(\Omega)} = 2(\mathcal{J}(w) - \mathcal{J}(u)) \;\; or \;\; \mathcal{J}(u) \leq \mathcal{J}(w), \tag{4.15}
$$

where Eq. 4.15 is known as the Principle of Minimum Potential Energy (PMPE). In other words, the true solution possesses the minimum potential.

The minimum property of the exact solution can be proven by an alternative technique. Let us construct a potential function, for a deviation away from the exact solution u, denoted $u + \lambda\nu$, where λ is a scalar and ν is any admissible variation (test function)

$$
\mathcal{J}(u+\lambda\nu) = \int_{\Omega} \frac{1}{2} \frac{d}{dx}(u+\lambda\nu) E \frac{d}{dx}(u+\lambda\nu)\, dx - \int_{\Omega} f(u+\lambda\nu)\, dx - t^{*}(u+\lambda\nu)|_{\Gamma_t}. \tag{4.16}
$$

If we differentiate with respect to λ,

$$
\frac{\partial \mathcal{J}(u+\lambda\nu)}{\partial\lambda} = \int_{\Omega} \frac{d\nu}{dx} E \frac{d}{dx}(u+\lambda\nu)\, dx - \int_{\Omega} f\nu\, dx - t^{*}\nu|_{\Gamma_t}, \tag{4.17}
$$

and set $\lambda = 0$ (because we know that the exact solution is for $\lambda = 0$), we have

$$
\frac{\partial \mathcal{J}(u+\lambda\nu)}{\partial\lambda}|_{\lambda=0} = \int_{\Omega} \frac{d\nu}{dx} E \frac{du}{dx}\, dx - \int_{\Omega} f\nu\, dx - t^{*}\nu|_{\Gamma_t} = 0. \tag{4.18}
$$

Clearly, the minimizer of the potential is the solution to the field equations, since it produces the weak formulation as a result. This is a minimum since

$$
\frac{\partial^2 \mathcal{J}(u+\lambda\nu)}{\partial\lambda^2}|_{\lambda=0} = \int_{\Omega} \frac{d\nu}{dx} E \frac{d\nu}{dx}\, dx \geq 0. \tag{4.19}
$$

It is important to note that the weak form, derived earlier, requires no such potential and thus is a more general approach than a minimum principle. Thus, in cases where a potential energy exists, the weak formulation can be considered as a minimization of it. Numerical approaches based on this idea are usually referred to as Rayleigh–Ritz methods. This concept allows one to construct simple error estimates for global mesh refinement.

4.4 Simple Estimates for Adequate FEM Meshes

The previous results generate estimates for the mesh fineness for a desired accuracy. As stated earlier, under standard assumptions the classical a priori error estimate for the finite element method is (Eq. 4.3), $||u - u^h||_{E(\Omega)} \leq C(u, p)\, h^{min(r-1,p)} \stackrel{\text{def}}{=} \gamma$. Using the PMPE for a finite element solution (Eq. 4.15), with $w = u^h$, we have

$$||u - u^h||^2_{E(\Omega)} = 2(\mathcal{J}(u^h) - \mathcal{J}(u)). \tag{4.20}$$

By solving the associated boundary value problem for two successively finer meshes, $h_1 > h_2$, with the following property $\mathcal{J}(u^{h_1}) \geq \mathcal{J}(u^{h_2}) \geq \mathcal{J}(u^{h=0})$, we can set up the following system of equations for unknown constant C:

$$\begin{aligned} ||u - u^{h_1}||^2_{E(\Omega)} &= 2(\mathcal{J}(u^{h_1}) - \mathcal{J}(u)) \approx C^2 h_1^{2\gamma}, \\ ||u - u^{h_2}||^2_{E(\Omega)} &= 2(\mathcal{J}(u^{h_2}) - \mathcal{J}(u)) \approx C^2 h_2^{2\gamma}. \end{aligned} \tag{4.21}$$

Solving for C

$$C = \sqrt{\frac{2(\mathcal{J}(u^{h_1}) - \mathcal{J}(u^{h_2}))}{h_1^{2\gamma} - h_2^{2\gamma}}}. \tag{4.22}$$

One can now solve for the appropriate mesh size by writing

$$C h^{\gamma}_{tol} \approx TOL \Rightarrow h_{tol} \approx \left(\frac{TOL}{C}\right)^{\frac{1}{\gamma}}. \tag{4.23}$$

In summary, to monitor the discretization error, we apply the following (Fig. 4.2) algorithm ($K = 0.5$)

$$\boxed{\begin{array}{l} \textbf{STEP 1}: \text{SOLVE WITH COARSE MESH} = \text{h}_1 \Rightarrow u^{h_1} \Rightarrow \mathcal{J}(u^{h_1}) \\ \textbf{STEP 2}: \text{SOLVE WITH FINER MESH} = \text{h}_2 = \text{K} \times \text{h}_1 \Rightarrow u^{h_2} \Rightarrow \mathcal{J}(u^{h_2}) \\ \textbf{STEP 3}: \text{COMPUTE C} \Rightarrow h_{tol} \approx \left(\frac{TOL}{C}\right)^{\frac{1}{\gamma}}. \end{array}} \tag{4.24}$$

Remarks: While this scheme provides a simple estimate for the global mesh fineness needed, the meshes need to be locally refined to ensure tolerable accuracy throughout the domain.

Fig. 4.2 Successively refined (halved/embedded) meshes used to estimate the error

4.5 Local Mesh Refinement

Probably the simplest approach to local mesh refinement is to use the residual as a guide. Residual methods require no a posteriori system of equations to be solved. Such methods bound the error by making use of

- the FEM solution itself,
- the data on the boundary,
- the error equation and
- the Galerkin orthogonality property.

The approach is to form the following bound

$$||u-u^h||^2_{E(\Omega)} \leq C_1 \underbrace{\sum_{e=1}^{N} h_e^2 ||r_1||^2_{L^2(\Omega_e)}}_{interior} + C_2 \underbrace{\sum_{I=1}^{INT} h_{eI} ||[\![r_2]\!]||^2_{L^2(\partial\Omega_I)}}_{interfaces} + C_3 \underbrace{\sum_{J=1}^{B-INT} h_{eJ} ||r_3||^2_{L^2(\partial\Omega_{JB})}}_{exterior-boundary}, \tag{4.25}$$

where

- C_1, C_2, and C_3 are constants,
- h_e are the sizes of the element,
- the interior element residual is $r_1 = \frac{d}{dx}(E(x)\frac{du^h}{dx}) + f$,
- the interior interface "jump" residual is $[\![r_2]\!] = (E(x)\frac{du^h}{dx})_{x^+} - (E(x)\frac{du^h}{dx})_{x^-}$,
- the boundary interface ("dissatisfaction") residual is $r_3 = (E(x)\frac{du^h}{dx}) - t^*$ on Γ_t, and
- local error indicators are defined by

$$\zeta_e^2 \stackrel{\text{def}}{=} C_1 h_e^2 ||r_1||^2_{L^2(\Omega_e)} + C_2 h_{eI} ||\, [\![r_2]\!] \,||^2_{L^2(\partial\Omega_I)} + C_3 h_{eJ} ||r_3||^2_{L^2(\partial\Omega_{JB})}. \tag{4.26}$$

The local quantities ζ_e are used to decide whether an element is to be refined (Fig. 4.3). If $\zeta_e > TOL$, then the element is refined. Such estimates, used to guide local adaptive finite element mesh refinement techniques, were first developed in Babúska

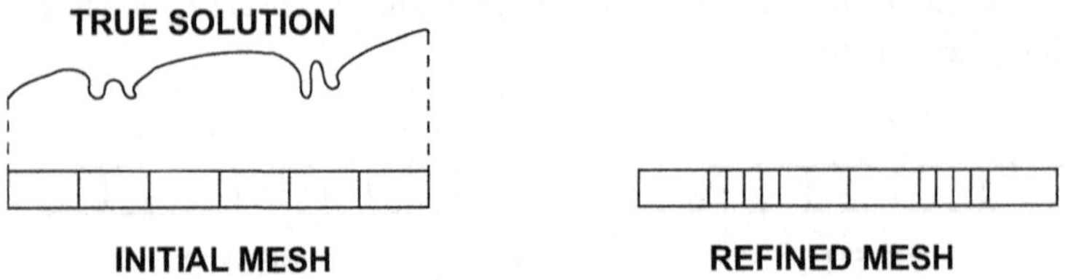

Fig. 4.3 Locally refined mesh to capture finer solution features

and Rheinboldt [9] for one-dimensional problems and in Babùska and Miller [10] and Kelly et al. [11] for two-dimensional problems. For reviews see Ainsworth and Oden [1]. This will be discussed further at the end of this monograph.

References

1. Ainsworth, M., & Oden, J. T. (2000). *A posterori error estimation in finite element analysis*. New York: Wiley.
2. Becker, E. B., Carey, G. F., & Oden, J. T. (1980). *Finite elements: An introduction*. Englewood Cliffs: Prentice-Hall.
3. Carey, G. F., & Oden, J. T. (1983). *Finite elements: A second course*. Englewood Cliffs: Prentice-Hall.
4. Oden, J. T., & Carey, G. F. (1984). *Finite elements: Mathematical aspects*. Englewood Cliffs: Prentice-Hall.
5. Hughes, T. J. R. (1989). *The finite element method*. Englewood Cliffs: Prentice Hall.
6. Szabo, B., & Babúska, I. (1991). *Finite element analysis*. New York: Wiley Interscience.
7. Bathe, K. J. (1996). *Finite element procedures*. Englewood Cliffs: Prentice-Hall.
8. Oden, J. T., & Demkowicz, L. F. (2010). *Applied functional analysis*. Boca Raton: CRC Press.
9. Babúska, I., & Rheinbolt, W. C. (1978). A posteriori error estimates for the finite element method. *The International Journal for Numerical Methods in Engineering*, *12*, 1597–1615.
10. Babúska, I., & Miller, A. D. (1987). A feedback finite element method with a-posteriori error estimation. Part I. *Computer Methods in Applied Mechanics and Engineering*, *61*, 1–40.
11. Kelly, D. W., Gago, J. R., Zienkiewicz, O. C., & Babùska, I. (1983). A posteriori error analysis and adaptive processes in the finite element method. Part I-error analysis. *International Journal for Numerical Methods in Engineering*, *19*, 1593–1619.

Iterative Solutions Schemes

5

5.1 Introduction: Minimum Principles and Krylov Methods

5.1.1 Numerical Linear Algebra

There are two main approaches to solving systems of equations resulting from numerical discretization of solid mechanics problems, direct and iterative. There are a large number of variants of each. Standard direct solvers are usually employed when the number of unknowns is not very large, and there are multiple load vectors.[1] Basically, one can operate on the multiple right-hand sides simultaneously via Gaussian elimination. For a back substitution the cost is

$$2(0+1+2+3...+N-1) = 2\sum_{k=1}^{N}(k-1) = N(N-1). \qquad (5.1)$$

Therefore the total cost of solving such a system is the cost to reduce the system to upper triangular form plus the cost of back substitution, i.e., $\frac{2}{3}N^3 + N(N-1)$. However since the operation counts to factor and solve an $N \times N$ system are $\mathcal{O}(N^3)$, iterative solvers are preferred when the systems are very large.[2] In general, most modern solvers, for large symmetric systems, like the ones of interest here, employ Conjugate Gradient (CG) type iterative techniques which can deliver solutions in

[1]However, specialized direct sparse solvers can be used if the matrices have a special structure.

[2]For example, Gaussian elimination is approximately

$$2(N+(N-1)^2+(N-2)^2+(N-3)^2+...1^2) = 2\sum_{k=1}^{N}k^2 \approx 2\int_0^N k^2dk = \frac{2}{3}N^3. \qquad (5.2)$$

An operation such as the addition of two numbers, or multiplication of two numbers, is one operation count.

T. I. Zohdi, *A Finite Element Primer for Beginners*, The Basics,
https://doi.org/10.1007/978-3-319-70428-9_5

$\mathcal{O}(N)^2$ operations.[3] It is inescapable, for almost all variants of Gaussian elimination, unless they involve complicated sparsity tracking to eliminate unneeded operations on zero entries that the operation costs are $\mathcal{O}(N^3)$. However band solvers, which exploit the band structure of the stiffness matrix, can reduce the number of operation counts to Nb^2, where b is the bandwidth. Many schemes exist for the optimal ordering of nodes in order to make the bandwidth as small as possible. Skyline solvers locate the uppermost nonzero elements starting from the diagonal and concentrate only on elements below the skyline. Frontal methods, which are analogous to a moving front in the finite element mesh, perform Gaussian elimination element by element, before the element is incorporated into the global stiffness matrix. In this procedure memory is reduced, and the elimination process can be done for all elements simultaneously, at least in theory. Upon assembly of the stiffness matrix and right-hand side, back substitution can be started immediately. If the operations are performed in an optimal order, it can be shown that the number of operations behaves proportionally to N^2. Such a process, is, of course, nontrivial. We note that with direct methods, zeros within the band, below the skyline, and in the front are generally filled and must be carried in the operations. In very large problems the storage requirements and the number of operation counts can become so large that solution by direct methods is not feasible. The data structures and I/O are also nontrivial concerns. However, we notice that a matrix/vector multiplication involves $2N^2$ operation counts, and that a method based on repeated vector multiplication, if convergent in less than N iterations, could be very attractive. This is usually the premise in using iterative methods, such as the Conjugate Gradient (CG) Method. A very important feature of iterative methods is that the memory requirements remain constant during the solution process. It is important to note that modern computer architectures are based on (1) *registers*, which have virtually no memory capabilities, but which can perform very fast operations on computer "words", (2) *cache*, which have slightly larger memory capabilities, with a slight reduction in speed, but are thermally very "hot" and are thus limited for physical as well as manufacturing reasons, (3) *main memory*, which is slower since I/O (input/output) is required, but still within a workstation, and (4) *disk and tape or magnetic drums*, which are out of the core system and thus require a huge I/O component and are very slow. Therefore, one point that we emphasize is that one can take advantage of the element-by-element structure inherent in the finite element method for data storage and matrix-vector multiplication in the CG method. The element-by-element data structure is also critical for the ability to fit matrix/vector multiplications into the computer cache, which essentially is a low memory/high floating point operation per second portion of the computer hardware.

Remark: *One singularly distinguishing feature of iterative solvers is the fact that since they are based on successive updates of a starting guess solution vector, they can be given a tremendous head start by a good solution guess, for example, provided*

[3]Similar iterative solvers can be developed for unsymmetric systems, and we refer the reader to Axelsson [1] for details.

by an analytical or semi-analytical solution. Minimum principles play a key role in the construction of a certain class of iterative solvers, which we exploit in the chapter.

5.1.2 Krylov Searches and Minimum Principles

By itself, the PMPE (introduced in the previous section) is a powerful theoretical result. However, it can be used to develop methods to solve systems of equations arising from a finite element discretization of a infinitesimal strain linearly elastic structure. This result is the essence of the so-called Krylov family of searches. Suppose we wish to solve the discrete system

$$[K]\{a\} = \{R\}. \tag{5.3}$$

$[K]$ is a symmetric positive definite $N \times N$ matrix; $\{a\}$ is the $N \times 1$ numerical solution vector, and $\{R\}$ is the $N \times 1$ right-hand side. We define a potential

$$\Pi \stackrel{\text{def}}{=} \frac{1}{2}\{a\}^T[K]\{a\} - \{a\}^T\{R\}. \tag{5.4}$$

Correspondingly, from basic calculus we have (see Appendix A)

$$\nabla\Pi \stackrel{\text{def}}{=} \{\frac{\partial\Pi}{\partial a_1}, \frac{\partial\Pi}{\partial a_2}, \ldots \frac{\partial\Pi}{\partial a_N}\}^T = 0 \Rightarrow [K]\{a\} - \{R\} = 0. \tag{5.5}$$

Therefore the minimizer of the potential Π is also the solution to the discrete system. A family of iterative solving techniques for symmetric systems based upon minimizing Π by successively updating a starting vector are the Krylov class. The minimization takes place over vector spaces called the Krylov spaces. These methods are based on the assumption that a solution, to a tolerable accuracy, can be achieved in much less than $\mathcal{O}(N^3)$ operations, as required with most Gaussian-type techniques. The simplest of this family is the method of steepest descent, which is a precursor to the widely used Conjugate Gradient Method.

5.1.2.1 The Method of Steepest Descent

The method of steepest descent is based upon the following simple idea: if the gradient of the potential is not zero at a possible solution vector, then the greatest increase of the scalar function is in the direction of the gradient; therefore we move in the opposite direction $-\nabla\Pi$. The ingredients in the methods are the residual,

$$\{r\}^i \stackrel{\text{def}}{=} -\nabla\Pi = \{R\} - [K]\{a\}^i, \tag{5.6}$$

and the successive iterates,

$$\{a\}^{i+1} \stackrel{\text{def}}{=} \{a\}^i + \lambda^i\{r\}^i. \tag{5.7}$$

We seek a λ^i such that Π is a global minimum. Directly we have

$$\Pi = \frac{1}{2}\{a\}^{T,i}[K]\{a\}^i + \lambda^i\{a\}^{T,i}[K]\{r\}^i + \frac{1}{2}\lambda^{i2}\{r\}^{T,i}[K]\{r\}^i - \{a\}^{T,i}\{R\} - \lambda^i\{r\}^{T,i}\{R\}, \tag{5.8}$$

where it was assumed that $[K]$ was symmetric. Forcing $\frac{\partial \Pi}{\partial \lambda^i} = 0$ and solving for λ^i yields

$$\lambda^i = \frac{\{r\}^{T,i}(\{R\} - [K]\{a\}^i)}{\{r\}^{T,i}[K]\{r\}^i} = \frac{\{r\}^{T,i}\{r\}^i}{\{r\}^{T,i}[K]\{r\}^i}. \tag{5.9}$$

Therefore the method of steepest descent consists of the following:

STEP 1 : SELECT A STARTING GUESS $\{a\}^1$

STEP 2 : COMPUTE :

$\{r\}^i = \{R\} - [K]\{a\}^i \qquad \lambda^i = \frac{\{r\}^{T,i}\{r\}^i}{\{r\}^{T,i}[K]\{r\}^i} \qquad \{a\}^{i+1} = \{a\}^i + \lambda^i\{r\}^i$

STEP 3 : COMPUTE :

$||\{a\}^{i+1} - \{a\}^i||_K^2 \stackrel{\text{def}}{=} (\{a\}^{T,i+1} - \{a\}^{T,i})[K](\{a\}^{i+1} - \{a\}^i) = \lambda^i ||\{r\}^i||_K^2$

If $||\{a\}^{i+1} - \{a\}^i||_K < \tau = \text{TOL} \Rightarrow$ STOP

If $||\{a\}^{i+1} - \{a\}^i||_K \geq \tau = \text{TOL} \Rightarrow$ GO TO STEP 2 WITH $i = i + 1$

The rate of convergence of the method is related to the condition number of the stiffness matrix

$$||\{a\} - \{a\}^i||_K = (1 - \frac{1}{C([K])})^{i/2}||\{a\} - \{a\}^1||_K, \tag{5.10}$$

where, in this case, the *Condition Number* is defined by

$$C([K]) \stackrel{\text{def}}{=} \frac{\text{max [K] eigenvalue}}{\text{min [K] eigenvalue}} \tag{5.11}$$

and $\{a\}$ is the exact solution to the algebraic system $[K]\{a\} = \{R\}$. The rate of convergence of the method is typically quite slow; however, a variant, the Conjugate Gradient Method, is guaranteed to converge in N iterations at most, provided the algebra is performed exactly.

5.1.2.2 The Conjugate Gradient Method

In the Conjugate Gradient Method, at each iteration the computational cost is $\mathcal{O}(N)$, due to the FEM matrix structure. We refer the reader to Axelsson [1] for details. We define the (matrix) residual,

$$\{r\}^i \stackrel{\text{def}}{=} -\nabla\Pi = \{R\} - [K]\{a\}^i, \tag{5.12}$$

and the successive iterates, for $i = 1, 2, 3...$,

$$\{a\}^{i+1} \stackrel{\text{def}}{=} \{a\}^i + \lambda^i \{z\}^i, \tag{5.13}$$

with

$$\{z\}^i \stackrel{\text{def}}{=} \{r\}^i + \theta^i \{z\}^{i-1}. \tag{5.14}$$

The coefficient θ^i is chosen so that $\{z\}^i$ is $[K] - conjugate$ to $\{z\}^{i-1}$, i.e.,

$$\{z\}^{T,i}[K]\{z\}^{i-1} = 0 \Rightarrow \theta^i = -\frac{\{r\}^{T,i}[K]\{z\}^{i-1}}{\{z\}^{T,i-1}[K]\{z\}^{i-1}}. \tag{5.15}$$

The value of λ^i which minimizes

$$\Pi = \frac{1}{2}(\{a\}^i + \lambda^i\{z\}^i)^T[K](\{a\}^i + \lambda^i\{z\}^i) - (\{a\}^i + \lambda^i\{z\}^i)^T\{R\}, \tag{5.16}$$

is (for $i = 1, 2, 3...$),

$$\lambda^i = \frac{\{z\}^{T,i}(\{R\} - [K]\{a\}^i)}{\{z\}^{T,i}[K]\{z\}^i} = \frac{\{z\}^{T,i}\{r\}^i}{\{z\}^{T,i}[K]\{z\}^i}. \tag{5.17}$$

The solution steps are:

STEP 1 : FOR i = 1 :$SELECT$ $\{a\}^1 \Rightarrow \{r\}^1 = \{R\} - [K]\{a\}^1 = \{z\}^1$

STEP 2 : COMPUTE (WITH $\{z\}^1 = \{r\}^1$)

$$\lambda^1 = \frac{\{z\}^{T,1}(\{R\}-[K]\{a\}^1)}{\{z\}^{T,1}[K]\{z\}^1} = \frac{\{z\}^{T,1}\{r\}^1}{\{z\}^{T,1}[K]\{z\}^1}$$

STEP 3 : COMPUTE $\{a\}^2 = \{a\}^1 + \lambda^1\{z\}^1$

STEP 4 : (FOR i > 1) COMPUTE $\{r\}^i = \{R\} - [K]\{a\}^i$

$$\theta^i = -\frac{\{r\}^{T,i}[K]\{z\}^{i-1}}{\{z\}^{T,i-1}[K]\{z\}^{i-1}} \qquad \{z\}^i \stackrel{\text{def}}{=} \{r\}^i + \theta^i\{z\}^{i-1} \tag{5.18}$$

$$\lambda^i = \frac{\{z\}^{T,i}(\{R\}-[K]\{a\}^i)}{\{z\}^{T,i}[K]\{z\}^i} = \frac{\{z\}^{T,i}\{r\}^i}{\{z\}^{T,i}[K]\{z\}^i}$$

COMPUTE $\{a\}^{i+1} = \{a\}^i + \lambda^i\{z\}^i$

STEP 5 : COMPUTE $e^i \stackrel{\text{def}}{=} \frac{||\{a\}^{i+1}-\{a\}^i||_K}{||\{a\}^i||_K} = \frac{|\lambda^i|||\{z\}^i||_K}{||\{a\}^i||_K} \leq \tau \ (\tau = TOL)$

IF $e^i < \tau \Rightarrow$ STOP

IF $e^i \geq \tau \Rightarrow$ GO TO STEP 4 AND REPEAT.

5.1.2.3 Accelerating Computations

The rate of convergence of the CG method is related to the condition number

$$||\{a\} - \{a\}^i||_K \leq (\frac{\sqrt{C([K])} - 1}{\sqrt{C([K])} + 1})^i ||\{a\} - \{a\}^1||_K. \tag{5.19}$$

Proofs of the various characteristics of the method can be found in Axelsson [1]. As is standard, in an attempt to reduce the condition number and hence increase the rate of convergence, typically preconditioning of $[K]$ is done by forming the following transformation of variables, $\{a\} = [T]\{\hat{a}\}$, which produces a preconditioned system, with stiffness matrix $\overline{[K]} = [T]^T[K][T]$. Ideally we would like $[T] = [L]^{-T}$ where $[L][L]^T = [K]$, and where $[L]$ is a lower triangular matrix, thus forcing

$$[T]^T[K][T] = [L]^{-1}[L][L]^T[L]^{-T} = I. \tag{5.20}$$

However, the reduction of the stiffness matrix into a lower triangular matrix and its transpose is comparable in the number of operations to solving the system by Gaussian elimination. Therefore, only an approximation to $[L]^{-1}$ is computed. Thus inexpensive preconditioners are usually used. For example, diagonal preconditioning, which is essentially the least expensive, involves defining $[T]$ as a diagonal matrix with entries

$$T_{ii} = \frac{1}{\sqrt{K_{ii}}}, i, j = 1, ...N, \tag{5.21}$$

where $T_{ij} = 0$ for $i \neq j$ and where the K_{ii} (no implied sum on the repeated indices) are the diagonal entries of $[K]$. In this case the resulting terms in the preconditioned stiffness matrix are unity on the diagonal. The off-diagonal terms, K_{ij}, are divided by $\frac{1}{\sqrt{K_{ii}}\sqrt{K_{jj}}}$. There are a variety of other preconditioning techniques, of widely ranging expense to compute. For more details see Axelsson [1]. It is strongly suggested to precondition the system. For example, with the simple diagonal preconditioner we obtain the following stiffness matrix

$$\begin{bmatrix}
1 & \frac{K_{12}}{\sqrt{K_{11}}\sqrt{K_{22}}} & \frac{K_{13}}{\sqrt{K_{11}}\sqrt{K_{33}}} & \frac{K_{14}}{\sqrt{K_{11}}\sqrt{K_{44}}} & \frac{K_{15}}{\sqrt{K_{11}}\sqrt{K_{55}}} & \cdot\,\cdot\,\ldots \\
\frac{K_{21}}{\sqrt{K_{11}}\sqrt{K_{22}}} & 1 & \frac{K_{23}}{\sqrt{K_{22}}\sqrt{K_{33}}} & \frac{K_{24}}{\sqrt{K_{22}}\sqrt{K_{44}}} & \cdot & \cdot\,\cdot\,\ldots \\
\frac{K_{31}}{\sqrt{K_{33}}\sqrt{K_{11}}} & \frac{K_{32}}{\sqrt{K_{33}}\sqrt{K_{22}}} & 1 & \cdot & \cdot & \cdot\,\cdot\,\ldots \\
\frac{K_{41}}{\sqrt{K_{44}}\sqrt{K_{11}}} & \frac{K_{42}}{\sqrt{K_{44}}\sqrt{K_{22}}} & \cdot & 1 & \cdot & \cdot\,\cdot\,\ldots \\
\frac{K_{51}}{\sqrt{K_{55}}\sqrt{K_{11}}} & \cdot & \cdot & \cdot & 1 & \cdot\,\cdot\,\ldots \\
\cdot & \cdot & \cdot & \cdot & \cdot & \cdot\,1\,\ldots \\
\cdot & \cdot & \cdot & \cdot & \cdot & \cdot\,\cdot\,\ldots \\
\cdot & \cdot & \cdot & \cdot & \cdot & \cdot\,\cdot\,\ldots
\end{bmatrix}. \tag{5.22}$$

Remark: In the one-dimensional problem considered earlier, the actual computation cost of the matrix-vector multiplication in an element-by-element CG method is a $[2 \times 2]$ matrix times a $\{2 \times 1\}$ vector times the number of elements. This is an $\mathcal{O}(N)$ calculation. If we consider M iterations necessary for convergence below an error tolerance, then the entire operation costs $\mathcal{O}(MN)$.

Reference

1. Axelsson, O. (1994). *Iterative solution methods*. Cambridge: Cambridge University Press.

Weak Formulations in Three Dimensions

6

6.1 Introduction

Albeit a bit repetitive, we follow similar constructions as done in the one-dimensional analysis of the preceding chapters. This allows readers a chance to contrast and compare the differences between one-dimensional and three-dimensional formulations. To derive a direct weak form for a body, we take the balance of linear momentum $\nabla \cdot \boldsymbol{\sigma} + \boldsymbol{f} = \mathbf{0}$ (denoting the strong form) and form a scalar product with an arbitrary smooth vector-valued function $\boldsymbol{\nu}$, and integrate over the body (Fig. 6.1),

$$\int_{\Omega} (\nabla \cdot \boldsymbol{\sigma} + \boldsymbol{f}) \cdot \boldsymbol{\nu} \, d\Omega = \int_{\Omega} \boldsymbol{r} \cdot \boldsymbol{\nu} \, d\Omega = 0, \tag{6.1}$$

where $\boldsymbol{r}$ is the residual and $\boldsymbol{\nu}$ is a test function. If we were to add a condition that we do this for all possible test functions ($\forall \boldsymbol{\nu}$), Eq. 6.1 implies $\boldsymbol{r} = \mathbf{0}$. Therefore, if every possible test function was considered, then

$$\boldsymbol{r} = \nabla \cdot \boldsymbol{\sigma} + \boldsymbol{f} = \mathbf{0} \tag{6.2}$$

on any finite region in Ω. Consequently, the weak and strong statements would be equivalent provided the true solution is smooth enough to have a strong solution. Clearly, $\boldsymbol{r}$ can never be nonzero over any finite region in the body, because the test function will locate them. Using the product rule of differentiation,

$$\nabla \cdot (\boldsymbol{\sigma} \cdot \boldsymbol{\nu}) = (\nabla \cdot \boldsymbol{\sigma}) \cdot \boldsymbol{\nu} + \nabla \boldsymbol{\nu} : \boldsymbol{\sigma} \tag{6.3}$$

leads to, $\forall \boldsymbol{\nu}$

$$\int_{\Omega} (\nabla \cdot (\boldsymbol{\sigma} \cdot \boldsymbol{\nu}) - \nabla \boldsymbol{\nu} : \boldsymbol{\sigma}) \, d\Omega + \int_{\Omega} \boldsymbol{f} \cdot \boldsymbol{\nu} \, d\Omega = 0, \tag{6.4}$$

T. I. Zohdi, *A Finite Element Primer for Beginners*, The Basics,
https://doi.org/10.1007/978-3-319-70428-9_6

Fig. 6.1 A three-dimensional body

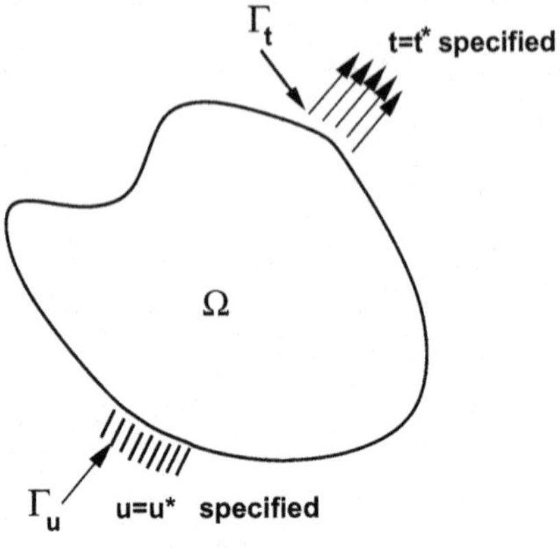

where we choose the $\boldsymbol{\nu}$ from an admissible set, to be discussed momentarily. Using the divergence theorem leads to, $\forall \boldsymbol{\nu}$,

$$\int_{\Omega} \nabla \boldsymbol{\nu} : \boldsymbol{\sigma} \, d\Omega = \int_{\Omega} \boldsymbol{f} \cdot \boldsymbol{\nu} \, d\Omega + \int_{\partial\Omega} \boldsymbol{\sigma} \cdot \boldsymbol{n} \cdot \boldsymbol{\nu} \, dA, \tag{6.5}$$

which, since the traction $\boldsymbol{t} = \boldsymbol{\sigma} \cdot \boldsymbol{n}$, leads to

$$\int_{\Omega} \nabla \boldsymbol{\nu} : \boldsymbol{\sigma} \, d\Omega = \int_{\Omega} \boldsymbol{f} \cdot \boldsymbol{\nu} \, d\Omega + \int_{\Gamma_t} \boldsymbol{t} \cdot \boldsymbol{\nu} \, dA. \tag{6.6}$$

If we decide to restrict our choices of $\boldsymbol{\nu}$'s to those such that $\boldsymbol{\nu}|_{\Gamma_u} = \boldsymbol{0}$, we have, where $\boldsymbol{u}^*$ is the applied boundary displacement on Γ_u, for infinitesimal strain linear elasticity

$$\text{Find } \boldsymbol{u}, \boldsymbol{u}|_{\Gamma_u} = \boldsymbol{u}^*, \quad \text{such that } \forall \boldsymbol{\nu}, \boldsymbol{\nu}|_{\Gamma_u} = \boldsymbol{0}$$

$$\underbrace{\int_{\Omega} \nabla \boldsymbol{\nu} : \mathbb{E} : \nabla \boldsymbol{u} \, d\Omega}_{\stackrel{\text{def}}{=} \mathcal{B}(\boldsymbol{u},\boldsymbol{\nu})} = \underbrace{\int_{\Omega} \boldsymbol{f} \cdot \boldsymbol{\nu} \, d\Omega + \int_{\Gamma_t} \boldsymbol{t}^* \cdot \boldsymbol{\nu} \, dA}_{\stackrel{\text{def}}{=} \mathcal{F}(\boldsymbol{\nu})}, \tag{6.7}$$

where $\boldsymbol{t} = \boldsymbol{t}^*$ on Γ_t. As in the one-dimensional formulation, this is called a "weak" form because it does not require the differentiability of the stress $\boldsymbol{\sigma}$. In other words, the differentiability requirements have been *weakened*. It is clear that we are able to consider problems with quite irregular solutions. We emphasize that if we test the solution with all possible test functions of sufficient smoothness, then the weak solution is equivalent to the strong solution. *Futhermore, that provided the true solution is smooth enough, the weak and strong forms are equivalent, which can be seen by the above constructive derivation.*

6.2 Hilbertian Sobolev Spaces

As in one dimension, a key question is the selection of the sets of functions in the weak form. Somewhat naively, the answer is simple, the integrals must remain finite. Therefore the following restrictions hold $(\forall \boldsymbol{\nu})$, $\int_{\Omega} \boldsymbol{f} \cdot \boldsymbol{\nu}\, d\Omega < \infty$, $\int_{\partial\Omega} \boldsymbol{t}^* \cdot \boldsymbol{\nu}\, dA < \infty$ and $\int_{\Omega} \nabla \boldsymbol{\nu} : \boldsymbol{\sigma}\, d\Omega < \infty$, and govern the selection of the approximation spaces. These relations simply mean that the functions must be square integrable. In order to make precise statements one must have a method of "book keeping." Such a system is to employ so-called Hilbertian Sobolev spaces. We recall that a norm has three main characteristics for any functions $\boldsymbol{u}$ and $\boldsymbol{\nu}$ such that $||\boldsymbol{u}|| < \infty$ and $||\boldsymbol{\nu}|| < \infty$ are

- (1) $||\boldsymbol{u}|| > 0$, $||\boldsymbol{u}|| = 0$ if and only if $\boldsymbol{u} = \boldsymbol{0}$,
- (2) $||\boldsymbol{u} + \boldsymbol{\nu}|| \leq ||\boldsymbol{u}|| + ||\boldsymbol{\nu}||$ and
- (3) $||\alpha \boldsymbol{u}|| = |\alpha| ||\boldsymbol{u}||$,

where α is a scalar constant. Certain types of norms, so-called Hilbert space norms, are frequently used in solid mechanics. Following standard notation, we denote $H^1(\Omega)$ as the usual space of scalar functions with generalized partial derivatives of order ≤ 1 in $L^2(\Omega)$, i.e., square integrable, in other words $u \in H^1(\Omega)$ if

$$||u||^2_{H^1(\Omega)} \stackrel{\text{def}}{=} \int_{\Omega} \sum_{j=1}^{3} \frac{\partial u}{\partial x_j} \frac{\partial u}{\partial x_j}\, d\Omega + \int_{\Omega} u u\, d\Omega < \infty. \tag{6.8}$$

Similarly, we define $\boldsymbol{H}^1(\Omega) \stackrel{\text{def}}{=} [H^1(\Omega)]^3$ as the space of vector-valued functions whose components are in $H^1(\Omega)$, i.e.,

$$\boldsymbol{u} \in \boldsymbol{H}^1(\Omega) \text{ if } ||\boldsymbol{u}||^2_{\boldsymbol{H}^1(\Omega)} \stackrel{\text{def}}{=} \int_{\Omega} \sum_{j=1}^{3} \sum_{i=1}^{3} \frac{\partial u_i}{\partial x_j} \frac{\partial u_i}{\partial x_j}\, d\Omega + \int_{\Omega} \sum_{i=1}^{3} u_i u_i\, d\Omega < \infty, \tag{6.9}$$

and we denote $\boldsymbol{L}^2(\Omega) \stackrel{\text{def}}{=} [L^2(\Omega)]^3$. Using these definitions, a complete boundary value problem can be written as follows. The data (loads) are assumed to be such that $\boldsymbol{f} \in \boldsymbol{L}^2(\Omega)$ and $\boldsymbol{t}^* \in \boldsymbol{L}^2(\Gamma_t)$, but less smooth data can be considered without complications. Implicitly we require that $\boldsymbol{u} \in \boldsymbol{H}^1(\Omega)$ and $\boldsymbol{\sigma} \in \boldsymbol{L}^2(\Omega)$ without continually making such references. Therefore, in summary we assume that our solutions obey these restrictions, leading to the following infinitesimal strain linear elasticity weak form:

$$\boxed{\begin{array}{l} \text{Find } \boldsymbol{u} \in \boldsymbol{H}^1(\Omega), \boldsymbol{u}|_{\Gamma_u} = \boldsymbol{u}^*, \quad \text{such that } \forall \boldsymbol{\nu} \in \boldsymbol{H}^1(\Omega), \boldsymbol{\nu}|_{\Gamma_u} = \boldsymbol{0} \\ \displaystyle\int_{\Omega} \nabla \boldsymbol{\nu} : \mathbb{E} : \nabla \boldsymbol{u}\, d\Omega = \int_{\Omega} \boldsymbol{f} \cdot \boldsymbol{\nu}\, d\Omega + \int_{\Gamma_t} \boldsymbol{t}^* \cdot \boldsymbol{\nu}\, dA. \end{array}} \tag{6.10}$$

We note that if the data in (6.10) are smooth and if (6.10) possesses a solution $\boldsymbol{u}$ that is sufficiently regular, then $\boldsymbol{u}$ is the solution of the classical linear elastostatics problem in strong form:

$$\boxed{\begin{aligned} &\nabla\cdot(\mathbb{E}:\nabla\boldsymbol{u})+\boldsymbol{f}=\boldsymbol{0}, \quad \forall \boldsymbol{x}\in\Omega,\\ &\boldsymbol{u}=\boldsymbol{u}^*, \quad \forall\boldsymbol{x}\in\Gamma_u,\\ &\boldsymbol{\sigma}\cdot\boldsymbol{n}=(\mathbb{E}:\nabla\boldsymbol{u})\cdot\boldsymbol{n}=\boldsymbol{t}=\boldsymbol{t}^*, \quad \forall\boldsymbol{x}\in\Gamma_t.\end{aligned}} \tag{6.11}$$

6.3 The Principle of Minimum Potential Energy

Repeating the procedure that we performed for one-dimensional formulations earlier in the monograph, we have

$$\begin{aligned} ||\boldsymbol{u}-\boldsymbol{w}||^2_{E(\Omega)} &= \mathcal{B}(\boldsymbol{u}-\boldsymbol{w},\boldsymbol{u}-\boldsymbol{w})\\ &=\mathcal{B}(\boldsymbol{u},\boldsymbol{u})+\mathcal{B}(\boldsymbol{w},\boldsymbol{w})-2\mathcal{B}(\boldsymbol{u},\boldsymbol{w})\\ &=\mathcal{B}(\boldsymbol{w},\boldsymbol{w})-\mathcal{B}(\boldsymbol{u},\boldsymbol{u})-2\mathcal{B}(\boldsymbol{u},\boldsymbol{w})+2\mathcal{B}(\boldsymbol{u},\boldsymbol{u})\\ &=\mathcal{B}(\boldsymbol{w},\boldsymbol{w})-\mathcal{B}(\boldsymbol{u},\boldsymbol{u})-2\mathcal{B}(\boldsymbol{u},\boldsymbol{w}-\boldsymbol{u})\\ &=\mathcal{B}(\boldsymbol{w},\boldsymbol{w})-\mathcal{B}(\boldsymbol{u},\boldsymbol{u})-2\mathcal{F}(\boldsymbol{w}-\boldsymbol{u})\\ &=\mathcal{B}(\boldsymbol{w},\boldsymbol{w})-2\mathcal{F}(\boldsymbol{w})-(\mathcal{B}(\boldsymbol{u},\boldsymbol{u})-2\mathcal{F}(\boldsymbol{u}))\\ &=2\mathcal{J}(\boldsymbol{w})-2\mathcal{J}(\boldsymbol{u}), \end{aligned} \tag{6.12}$$

where similar to the one-dimensional case, we define the *elastic potential* as

$$\mathcal{J}(\boldsymbol{w}) \stackrel{\text{def}}{=} \frac{1}{2}\mathcal{B}(\boldsymbol{w},\boldsymbol{w})-\mathcal{F}(\boldsymbol{w})=\frac{1}{2}\int_\Omega \nabla\boldsymbol{w}:\mathbb{E}:\nabla\boldsymbol{w}\,d\Omega-\int_\Omega \boldsymbol{f}\cdot\boldsymbol{w}\,d\Omega-\int_{\Gamma_t}\boldsymbol{t}^*\cdot\boldsymbol{w}\,dA. \tag{6.13}$$

This implies

$$0\le ||\boldsymbol{u}-\boldsymbol{w}||^2_{E(\Omega)}=2(\mathcal{J}(\boldsymbol{w})-\mathcal{J}(\boldsymbol{u})) \;\; or \;\; \mathcal{J}(\boldsymbol{u})\le\mathcal{J}(\boldsymbol{w}), \tag{6.14}$$

where Eq. 6.14 is known as the Principle of Minimum Potential Energy (PMPE). In other words, the true solution possesses the minimum potential. As in one dimension, the minimum property of the exact solution can be proven by an alternative technique. Let us construct a potential function, for a deviation away from the exact solution $\boldsymbol{u}$, denoted $\boldsymbol{u}+\lambda\boldsymbol{\nu}$, where λ is a scalar and $\boldsymbol{\nu}$ is any admissible variation (test function)

$$\mathcal{J}(\boldsymbol{u}+\lambda\boldsymbol{\nu})=\frac{1}{2}\int_\Omega \nabla(\boldsymbol{u}+\lambda\boldsymbol{\nu}):\mathbb{E}:\nabla(\boldsymbol{u}+\lambda\boldsymbol{\nu})\,d\Omega-\int_\Omega\boldsymbol{f}\cdot(\boldsymbol{u}+\lambda\boldsymbol{\nu})\,d\Omega-\int_{\Gamma_t}\boldsymbol{t}^*\cdot(\boldsymbol{u}+\lambda\boldsymbol{\nu})\,dA. \tag{6.15}$$

If we differentiate with respect to λ,

$$\frac{\partial \mathcal{J}(\boldsymbol{u}+\lambda\boldsymbol{\nu})}{\partial \lambda} = \int_{\Omega} \nabla\boldsymbol{\nu} : \mathbb{E} : \nabla(\boldsymbol{u}+\lambda\boldsymbol{\nu})\, d\Omega - \int_{\Omega} \boldsymbol{f}\cdot\boldsymbol{\nu}\, d\Omega - \int_{\Gamma_t} \boldsymbol{t}^{*}\cdot\boldsymbol{\nu}\, dA, \tag{6.16}$$

and set $\lambda = 0$ (because we know that the exact solution is for $\lambda = 0$), we have

$$\frac{\partial \mathcal{J}(\boldsymbol{u}+\lambda\boldsymbol{\nu})}{\partial \lambda}|_{\lambda=0} = \int_{\Omega} \nabla\boldsymbol{\nu} : \mathbb{E} : \nabla\boldsymbol{u}\, d\Omega - \int_{\Omega} \boldsymbol{f}\cdot\boldsymbol{\nu}\, d\Omega - \int_{\Gamma_t} \boldsymbol{t}^{*}\cdot\boldsymbol{\nu}\, dA = 0. \tag{6.17}$$

Clearly, the minimizer of the potential is the solution to the field equations, since it produces the weak form as a result. This is a minimum since

$$\frac{\partial^2 \mathcal{J}(\boldsymbol{u}+\lambda\boldsymbol{\nu})}{\partial \lambda^2}|_{\lambda=0} = \int_{\Omega} \nabla\boldsymbol{\nu} : \mathbb{E} : \nabla\boldsymbol{\nu}\, d\Omega \geq 0. \tag{6.18}$$

It is important to note that the weak form, derived earlier, requires no such potential, and thus is a more general approach than a minimum principle. Thus, in the hyperelastic case, the weak formulation can be considered as a minimization of a potential energy function. This is sometimes referred to as the Rayleigh–Ritz method.

6.4 Complementary Principles

There exist another set of weak forms and minimum principles called complementary principles. Starting with $\nabla\cdot\boldsymbol{\tau} = \boldsymbol{0}$, $\boldsymbol{\tau}\cdot\boldsymbol{n}|_{\Gamma_t} = \boldsymbol{0}$, multiplying by the solution $\boldsymbol{u}$ leads to

$$\int_{\Omega} \nabla\cdot\boldsymbol{\tau}\cdot\boldsymbol{u}\, d\Omega = 0 = \int_{\Omega} \nabla\cdot(\boldsymbol{\tau}\cdot\boldsymbol{u})\, d\Omega - \int_{\Omega} \boldsymbol{\tau} : \nabla\boldsymbol{u}\, d\Omega. \tag{6.19}$$

Using the divergence theorem yields

$$\boxed{\begin{array}{l} \text{Find } \boldsymbol{\sigma}, \nabla\cdot\boldsymbol{\sigma}+\boldsymbol{f}=\boldsymbol{0}, \boldsymbol{\sigma}\cdot\boldsymbol{n}|_{\Gamma_t}=\boldsymbol{t} \text{ such that} \\[2ex] \underbrace{\int_{\Omega} \boldsymbol{\tau} : \mathbb{E}^{-1} : \boldsymbol{\sigma}\, d\Omega}_{\stackrel{\text{def}}{=}\mathcal{A}(\boldsymbol{\sigma},\boldsymbol{\tau})} = \underbrace{\int_{\Gamma_u} \boldsymbol{\tau}\cdot\boldsymbol{n}\cdot\boldsymbol{u}^{*}\, dA}_{\stackrel{\text{def}}{=}\mathcal{G}(\boldsymbol{\tau})} \qquad \forall\boldsymbol{\tau}, \nabla\cdot\boldsymbol{\tau}=\boldsymbol{0}, \boldsymbol{\tau}\cdot\boldsymbol{n}|_{\Gamma_t}=\boldsymbol{0}. \end{array}} \tag{6.20}$$

This is called the complementary form of Eq. 6.7. Similar restrictions are placed on the trial and test fields to force the integrals to make sense, i.e., to be finite. Similar boundedness restrictions control the choice of admissible complementary functions. In other words we assume that the solutions produce finite energy. *Despite the apparent simplicity of such principles they are rarely used in practical computations, directly in this form, because of the fact that it is somewhat difficult to find approximate functions,* $\boldsymbol{\sigma}$, that satisfy $\nabla \cdot \boldsymbol{\sigma} + \boldsymbol{f} = \mathbf{0}$. However, in closing, we provide some theoretical results. As in the primal case, a similar process is repeated using the complementary weak form. We define a complementary norm

$$0 \leq ||\boldsymbol{\sigma} - \boldsymbol{\gamma}||^2_{E^{-1}(\Omega)} \stackrel{\text{def}}{=} \int_{\Omega} (\boldsymbol{\sigma} - \boldsymbol{\gamma}) : \mathbb{E}^{-1} : (\boldsymbol{\sigma} - \boldsymbol{\gamma})\, d\Omega = \mathcal{A}(\boldsymbol{\sigma} - \boldsymbol{\gamma}, \boldsymbol{\sigma} - \boldsymbol{\gamma}). \tag{6.21}$$

Again, by direct manipulation, we have

$$\begin{aligned}
||\boldsymbol{\sigma} - \boldsymbol{\gamma}||^2_{E^{-1}(\Omega)} &= \mathcal{A}(\boldsymbol{\sigma} - \boldsymbol{\gamma}, \boldsymbol{\sigma} - \boldsymbol{\gamma}) \\
&= \mathcal{A}(\boldsymbol{\sigma}, \boldsymbol{\sigma}) + \mathcal{A}(\boldsymbol{\gamma}, \boldsymbol{\gamma}) - 2\mathcal{A}(\boldsymbol{\sigma}, \boldsymbol{\gamma}) \\
&= \mathcal{A}(\boldsymbol{\gamma}, \boldsymbol{\gamma}) - \mathcal{A}(\boldsymbol{\sigma}, \boldsymbol{\sigma}) - 2\mathcal{A}(\boldsymbol{\sigma}, \boldsymbol{\gamma}) + 2\mathcal{A}(\boldsymbol{\sigma}, \boldsymbol{\sigma}) \\
&= \mathcal{A}(\boldsymbol{\gamma}, \boldsymbol{\gamma}) - \mathcal{A}(\boldsymbol{\sigma}, \boldsymbol{\sigma}) - 2\mathcal{A}(\boldsymbol{\sigma}, \boldsymbol{\gamma} - \boldsymbol{\sigma}) \\
&= \mathcal{A}(\boldsymbol{\gamma}, \boldsymbol{\gamma}) - \mathcal{A}(\boldsymbol{\sigma}, \boldsymbol{\sigma}) - 2\mathcal{G}(\boldsymbol{\gamma} - \boldsymbol{\sigma}) \\
&= \mathcal{A}(\boldsymbol{\gamma}, \boldsymbol{\gamma}) - 2\mathcal{G}(\boldsymbol{\gamma}) - (\mathcal{A}(\boldsymbol{\sigma}, \boldsymbol{\sigma}) - 2\mathcal{G}(\boldsymbol{\sigma})) \\
&= 2\mathcal{K}(\boldsymbol{\gamma}) - 2\mathcal{K}(\boldsymbol{\sigma}),
\end{aligned} \tag{6.22}$$

where we define $\mathcal{K}(\boldsymbol{\gamma}) \stackrel{\text{def}}{=} \frac{1}{2}\mathcal{A}(\boldsymbol{\gamma}, \boldsymbol{\gamma}) - \mathcal{G}(\boldsymbol{\gamma}) = \frac{1}{2}\int_{\Omega} \boldsymbol{\gamma} : \mathbb{E}^{-1} : \boldsymbol{\gamma}\, d\Omega - \int_{\Gamma_u} \boldsymbol{\gamma} \cdot \boldsymbol{n} \cdot \boldsymbol{u}^*\, dA$. Therefore,

$$||\boldsymbol{\sigma} - \boldsymbol{\gamma}||^2_{E^{-1}(\Omega)} = 2(\mathcal{K}(\boldsymbol{\gamma}) - \mathcal{K}(\boldsymbol{\sigma})) \;\; or \;\; \mathcal{K}(\boldsymbol{\sigma}) \leq \mathcal{K}(\boldsymbol{\gamma}), \tag{6.23}$$

which is the Principle of Minimum Complementary Potential Energy (PMCPE). By directly adding together the potential energy and the complementary energy we obtain an equation of balance:

$$\begin{aligned}
\mathcal{J}(\boldsymbol{u}) + \mathcal{K}(\boldsymbol{\sigma}) &= \frac{1}{2}\int_{\Omega} \nabla \boldsymbol{u} : \mathbb{E} : \nabla \boldsymbol{u}\, d\Omega - \int_{\Omega} \boldsymbol{f} \cdot \boldsymbol{u}\, d\Omega - \int_{\Gamma_t} \boldsymbol{t}^* \cdot \boldsymbol{u}\, dA \\
&+ \frac{1}{2}\int_{\Omega} \boldsymbol{\sigma} : \mathbb{E}^{-1} : \boldsymbol{\sigma}\, d\Omega - \int_{\Gamma_u} \underbrace{\boldsymbol{t} \cdot \boldsymbol{u}}_{(\boldsymbol{\sigma} \cdot \boldsymbol{n}) \cdot \boldsymbol{u}^*}\, dA \\
&= 0.
\end{aligned} \tag{6.24}$$

Remark: Basically, the three-dimensional and one-dimensional formulations are, formally speaking, virtually identical in structure.

A Finite Element Implementation in Three Dimensions

7

7.1 Introduction

Generally, the ability to change the boundary data quickly is very important in finite element computations. One approach to do this rapidly is via the variational penalty method. This is done by relaxing kinematic assumptions on the members of the space of admissible functions and adding a term to "account for the violation" on the boundary. This is a widely used practice, and therefore to keep the formulation as general as possible we include penalty terms, *although this implementation is not mandatory*. Obviously, one could simply extract the known (imposed) values of boundary displacements (by appropriately eliminating rows and columns and modifying the right-hand side load vector); however, it is tedious. Nevertheless we consider the penalty method formulation for generality, although one does not necessarily need to use it. Accordingly, consider the following statement: Find $\boldsymbol{u} \in \boldsymbol{H}^1(\Omega)$ such that $\forall \boldsymbol{v} \in \boldsymbol{H}^1(\Omega)$

$$\int_{\Omega} \nabla \boldsymbol{v} : \mathbb{E} : \nabla \boldsymbol{u} \, d\Omega = \int_{\Omega} \boldsymbol{f} \cdot \boldsymbol{v} \, d\Omega + \int_{\Gamma_t} \boldsymbol{t}^* \cdot \boldsymbol{v} \, dA + P^\star \int_{\Gamma_u} (\boldsymbol{u}^* - \boldsymbol{u}) \cdot \boldsymbol{v} \, dA, \tag{7.1}$$

where the last term is to be thought of as a penalty term to enforce the applied displacement (kinematic) boundary condition, $\boldsymbol{u}|_{\Gamma_u} = \mathbf{u}^*$, and we relax the condition that the test function vanishes on the displacement part of the boundary, $\boldsymbol{v}|_{\Gamma_u} = \mathbf{0}$. The (penalty) parameter $P^\star$ is a large positive number. A penalty formulation has a variety of interpretations. It is probably simplest to interpret it as a traction that attempts to restore the correct prescribed displacement:

$$\int_{\Gamma_u} \boldsymbol{t} \cdot \boldsymbol{v} \, dA \approx P^\star \int_{\Gamma_u} (\boldsymbol{u}^* - \boldsymbol{u}) \cdot \boldsymbol{v} \, dA, \tag{7.2}$$

T. I. Zohdi, *A Finite Element Primer for Beginners*, The Basics,
https://doi.org/10.1007/978-3-319-70428-9_7

where the term $P^{\star}(\boldsymbol{u}^{*}-\boldsymbol{u})$ takes on the physical interpretation as a very stiff "traction spring" which is proportional to the amount of violation from the true boundary displacement.

Remark: In the case where a potential exists, as is the case here, we can motivate this approach by considering an augmented potential $\mathcal{J}(\boldsymbol{u}, P^{\star}) \stackrel{\text{def}}{=} \mathcal{J}(\boldsymbol{u}) + P^{\star}\int_{\Gamma_u}(\boldsymbol{u}^{*}-\boldsymbol{u})\cdot(\boldsymbol{u}^{*}-\boldsymbol{u})\,dA$, $\boldsymbol{u} \in \boldsymbol{H}^1(\Omega)$, whose variation is

Find $\boldsymbol{u} \in \boldsymbol{H}^1(\Omega)$ such that $\forall \boldsymbol{v} \in \boldsymbol{H}^1(\Omega)$

$$\int_{\Omega} \nabla\boldsymbol{v} : \mathbb{E} : \nabla\boldsymbol{u}\,d\Omega = \int_{\Omega} \boldsymbol{f}\cdot\boldsymbol{v}\,d\Omega + \int_{\Gamma_t} \boldsymbol{t}^{*}\cdot\boldsymbol{v}\,dA + P^{\star}\int_{\Gamma_u}(\boldsymbol{u}^{*}-\boldsymbol{u})\cdot\boldsymbol{v}\,dA. \tag{7.3}$$

Therefore, the penalty term can be thought of as a quadratic augmentation of the potential energy. When no potential exists, the penalty method can only be considered as an enforcement of a constraint.

7.2 FEM Approximation

It is convenient to write the bilinear form in the following (matrix) manner

$$\int_{\Omega} ([\boldsymbol{D}]\{\boldsymbol{v}\})^T[\mathbb{E}]([\boldsymbol{D}]\{\boldsymbol{u}\})\,d\Omega = \int_{\Omega}\{\boldsymbol{v}\}^T\{\boldsymbol{f}\}\,d\Omega + \int_{\Gamma_t}\{\boldsymbol{v}\}^T\{\boldsymbol{t}^{*}\}\,dA + P^{\star}\int_{\Gamma_u}\{\boldsymbol{v}\}^T\{\boldsymbol{u}^{*}-\boldsymbol{u}\}\,dA, \tag{7.4}$$

where $[\boldsymbol{D}]$, the deformation tensor, is

$$[\boldsymbol{D}] \stackrel{\text{def}}{=} \begin{bmatrix} \frac{\partial}{\partial x_1} & 0 & 0 \\ 0 & \frac{\partial}{\partial x_2} & 0 \\ 0 & 0 & \frac{\partial}{\partial x_3} \\ \frac{\partial}{\partial x_2} & \frac{\partial}{\partial x_1} & 0 \\ 0 & \frac{\partial}{\partial x_3} & \frac{\partial}{\partial x_2} \\ \frac{\partial}{\partial x_3} & 0 & \frac{\partial}{\partial x_1} \end{bmatrix}, \quad \{\boldsymbol{u}\} \stackrel{\text{def}}{=} \begin{Bmatrix} u_1 \\ u_2 \\ u_3 \end{Bmatrix}, \quad \{\boldsymbol{f}\} \stackrel{\text{def}}{=} \begin{Bmatrix} f_1 \\ f_2 \\ f_3 \end{Bmatrix}, \quad \{\boldsymbol{t}^{*}\} \stackrel{\text{def}}{=} \begin{Bmatrix} t_1^{*} \\ t_2^{*} \\ t_3^{*} \end{Bmatrix}. \tag{7.5}$$

It is clear that in an implementation of the finite element method, the sparsity of $[\boldsymbol{D}]$ should be taken into account. It is also convenient to write the approximations as

$$\boxed{\begin{aligned} u_1^h(x_1,x_2,x_3) &= \textstyle\sum_{i=1}^N a_i\phi_i(x_1,x_2,x_3),\\ u_2^h(x_1,x_2,x_3) &= \textstyle\sum_{i=1}^N a_{i+N}\phi_i(x_1,x_2,x_3),\\ u_3^h(x_1,x_2,x_3) &= \textstyle\sum_{i=1}^N a_{i+2N}\phi_i(x_1,x_2,x_3),\end{aligned}} \tag{7.6}$$

or $\{\boldsymbol{u}^h\} = [\phi]\{\boldsymbol{a}\}$, where, for example[1]:

$$[\phi] \stackrel{\text{def}}{=} \begin{bmatrix} \phi_1\,\phi_2\,\phi_3\,\phi_4\,\phi_5\,\phi_6 \ldots\phi_N & 0\;0\;0\;0\;0\;0\;0\;\ldots & 0\;0\;0\;0\;0\;0\;0\;\ldots \\ 0\;0\;0\;0\;0\;0\;0\;\ldots & \phi_1\,\phi_2\,\phi_3\,\phi_4\,\phi_5\,\phi_6 \ldots\phi_N & 0\;0\;0\;0\;0\;0\;0\;\ldots \\ 0\;0\;0\;0\;0\;0\;0\;\ldots & 0\;0\;0\;0\;0\;0\;0\;\ldots & \phi_1\,\phi_2\,\phi_3\,\phi_4\,\phi_5\,\phi_6 \ldots\phi_N \end{bmatrix}. \tag{7.7}$$

It is advantageous to write

$$\{\boldsymbol{a}\} \stackrel{\text{def}}{=} \left\{\begin{matrix} a_1\\ a_2\\ a_3\\ .\\ .\\ .\\ a_{3N}\end{matrix}\right\}, \quad \underbrace{\{\phi_i\} \stackrel{\text{def}}{=} \left\{\begin{matrix}\phi_i\\ 0\\ 0\end{matrix}\right\}}_{\text{for } 1\le i\le N}, \quad \underbrace{\{\phi_i\} \stackrel{\text{def}}{=} \left\{\begin{matrix}0\\ \phi_{i-N}\\ 0\end{matrix}\right\}}_{\text{for } N+1\le i\le 2N}, \quad \underbrace{\{\phi_i\} \stackrel{\text{def}}{=} \left\{\begin{matrix}0\\ 0\\ \phi_{i-2N}\end{matrix}\right\}}_{\text{for } 2N+1\le i\le 3N}, \tag{7.8}$$

and $\{\boldsymbol{u}^h\} = \sum_{i=1}^{3N} a_i\{\phi_i\}$. If we choose $\boldsymbol{v}$ with the same basis, but a different linear combination $\{\boldsymbol{v}^h\} = [\phi]\{\boldsymbol{b}\}$, then we may write

$$\begin{aligned}\underbrace{\int_\Omega ([\boldsymbol{D}][\phi]\{\boldsymbol{b}\})^T[\mathbb{E}]([\boldsymbol{D}][\phi]\{\boldsymbol{a}\})\,d\Omega}_{\{\boldsymbol{b}\}^T[\boldsymbol{K}]\{\text{stiffness}} &= \underbrace{\int_\Omega ([\phi]\{\boldsymbol{b}\})^T\{\boldsymbol{f}\}\,d\Omega}_{\text{body load}} + \underbrace{\int_{\Gamma_t} ([\phi]\{\boldsymbol{b}\})^T\{\boldsymbol{t}^*\}\,dA}_{\text{traction load}} \\ &\quad \underbrace{+P^\star \int_{\Gamma_u} ([\phi]\{\boldsymbol{b}\})^T\{\boldsymbol{u}^* - ([\phi]\{\boldsymbol{a}\})\}\,dA}_{\text{boundary penalty term}}\,.\end{aligned} \tag{7.9}$$

Since $\{\boldsymbol{b}\}$ is arbitrary ($\forall \boldsymbol{v} \Rightarrow \forall\{\boldsymbol{b}\}$), we have:

- $\{\boldsymbol{b}\}^T\{[\boldsymbol{K}]\{\boldsymbol{a}\} - \{\boldsymbol{R}\}\} = 0 \Rightarrow [\boldsymbol{K}]\{\boldsymbol{a}\} = \{\boldsymbol{R}\}$,
- $[\boldsymbol{K}] \stackrel{\text{def}}{=} \int_\Omega ([\boldsymbol{D}][\phi])^T[\mathbb{E}]([\boldsymbol{D}][\phi])\,d\Omega + P^\star \int_{\Gamma_u}[\phi]^T[\phi]\,dA$,
- $\{\boldsymbol{R}\} \stackrel{\text{def}}{=} \int_\Omega [\phi]^T\{\boldsymbol{f}\}\,d\Omega + \int_{\Gamma_t}[\phi]^T\{\boldsymbol{t}^*\}\,dA + P^\star \int_{\Gamma_u}[\phi]^T\{\boldsymbol{u}^*\}\,dA$.

Explicitly, $[\boldsymbol{K}]\{\boldsymbol{a}\} = \{\boldsymbol{R}\}$ is the system of equations that is to be solved.

---

[1]Representing the numerical approximation this way is simply to ease the understanding of the process. On the implementation level, one would not store the matrices in this form due to the large number of zeroes.

7.3 Global/Local Transformations

One strength of the finite element method is that most of the computations can be done in an element-by-element manner. We define the entries of $[\boldsymbol{K}]$,

$$[\boldsymbol{K}] = \int_{\Omega} ([\boldsymbol{D}][\phi])^T [\boldsymbol{I\!E}]([\boldsymbol{D}][\phi])\, d\Omega + P^{\star} \int_{\Gamma_u} [\phi]^T [\phi]\, dA \tag{7.10}$$

and

$$\{\boldsymbol{R}\} = \int_{\Omega} [\phi]^T \{\boldsymbol{f}\}\, d\Omega + \int_{\Gamma_t} [\phi]^T \{\boldsymbol{t}^*\}\, dA + P^{\star} \int_{\Gamma_u} [\phi]^T \{\boldsymbol{u}^*\}\, dA. \tag{7.11}$$

Breaking the calculations into elements, $[\boldsymbol{K}] = \sum_e [\boldsymbol{K}]^e$, $e = 1, 2, \ldots$ number of elements, where

$$[\boldsymbol{K}]^e = \int_{\Omega_e} ([\boldsymbol{D}][\phi])^T [\boldsymbol{I\!E}]([\boldsymbol{D}][\phi])\, d\Omega_e + P^{\star} \int_{\Gamma_{u,e}} [\phi]^T [\phi]\, dA_e, \tag{7.12}$$

and

$$\{\boldsymbol{R}\}^e = \int_{\Omega_e} [\phi]^T \{\boldsymbol{f}\}\, d\Omega_e + \int_{\Gamma_{t,e}} [\phi]^T \{\boldsymbol{t}^*\}\, dA_e + P^{\star} \int_{\Gamma_{u,e}} [\phi]^T \{\boldsymbol{u}^*\}\, dA_e, \tag{7.13}$$

where $\Gamma_{u,e} = \Gamma_u \cap \partial\Omega_e$ and $\Gamma_{t,e} = \Gamma_t \cap \partial\Omega_e$. In order to make the calculations systematic we wish to use the generic or master element defined in a local coordinate system $(\zeta_1, \zeta_2, \zeta_3)$. Accordingly, we need the following mapping functions (Fig. 7.1), from the master coordinates to the real space coordinates, $M_{x\zeta} : (\zeta_1, \zeta_2, \zeta_3) \mapsto (x_1, x_2, x_3)$ (e.g., trilinear bricks):

$$\boxed{\begin{aligned} x_1 &= \textstyle\sum_{i=1}^{8} \mathcal{X}_{1i} \hat{\phi}_i \stackrel{\text{def}}{=} M_{x\zeta_1}(\zeta_1, \zeta_2, \zeta_3), \\ x_2 &= \textstyle\sum_{i=1}^{8} \mathcal{X}_{2i} \hat{\phi}_i \stackrel{\text{def}}{=} M_{x\zeta_2}(\zeta_1, \zeta_2, \zeta_3), \\ x_3 &= \textstyle\sum_{i=1}^{8} \mathcal{X}_{3i} \hat{\phi}_i \stackrel{\text{def}}{=} M_{x\zeta_3}(\zeta_1, \zeta_2, \zeta_3), \end{aligned}} \tag{7.14}$$

where $(\mathcal{X}_{1i}, \mathcal{X}_{2i}, \mathcal{X}_{3i})$ are true spatial coordinates of the ith node and where $\hat{\phi}(\zeta_1, \zeta_2, \zeta_3) \stackrel{\text{def}}{=} \phi(x_1(\zeta_1, \zeta_2, \zeta_3), x_2(\zeta_1, \zeta_2, \zeta_3), x_3(\zeta_1, \zeta_2, \zeta_3))$. As in the one-dimension, these types of mappings are usually termed parametric maps. If the polynomial order of the shape functions is as high as the element, it is an isoparametric map, lower, then subparametric map, higher, then superparametric.

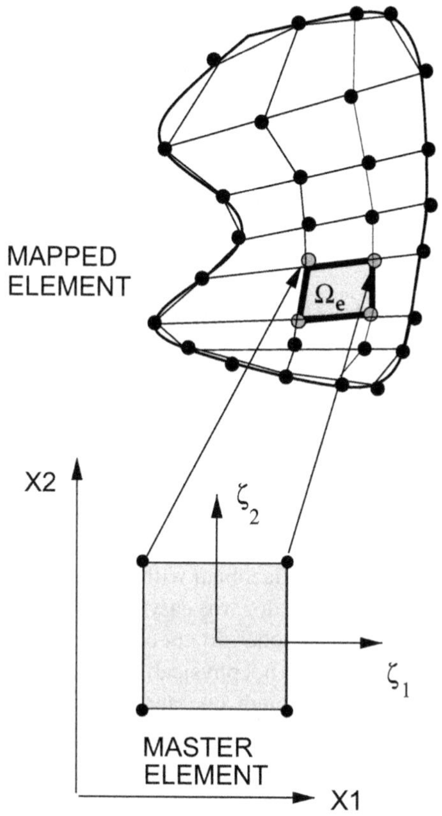

Fig. 7.1 A two-dimensional finite element mapping

7.4 Mesh Generation and Connectivity Functions

During the calculations, one needs to be able to connect the local numbering of the nodes to the global numbering of the nodes. For simple geometries, this is a straightforward scheme to automate. For complicated geometries, a lookup array connecting the local node number within an element to the global number is needed. Global/local connection is important since the proper (global) locations of the entries within the stiffness element are needed when solving the system of equations, either by Gaussian elimination or in element-by-element multiplication in a CG-type solver (Table 7.1).

Table 7.1 Local/global numbers for elements for an arch

Local node #	$e^{\#1}$ *node #*	$e^{\#2}$ *node #*	$e^{\#3}$ *node #*	$e^{\#4}$ *node #*
1	1	2	3	4
2	2	3	4	5
3	7	8	9	10
4	6	7	8	9

7.5 Warning: Restrictions on Elements

Recall that in one dimension we have the following type of calculation

$$\int_{x}^{x+h} \frac{d\phi_i}{dx} E \frac{d\phi_j}{dx}\, dx = \int_{-1}^{+1} \frac{d\hat{\phi}_i}{d\zeta}\frac{d\zeta}{dx} E \frac{d\hat{\phi}_j}{d\zeta}\frac{d\zeta}{dx}\frac{dx}{d\zeta}\, d\zeta = \int_{-1}^{+1} \frac{d\hat{\phi}_i}{d\zeta} E \frac{d\hat{\phi}_j}{d\zeta} \underbrace{\frac{d\zeta}{dx}}_{1/J}\, d\zeta. \tag{7.15}$$

Clearly, a zero Jacobian will lead to problems and potentially singular integrals. In one dimension, this was easy to avoid since the nodes are numbered sequentially and as long as the nodes do not coincide, this will not happen, since $J = h/2$. However, clearly, $J < 0$ is not physical, because this implies that neighboring nodes get mapped inside out (through one another). Negative Jacobians can also lead to indefinite stiffness matrices. As in one-dimensional formulations, for two-dimensional and three-dimensional formulations, one has to insure that $J = det\boldsymbol{F} > 0$, throughout the domain.

There are two ways that nonpositive Jacobians can occur: (1) The elements are nonconvex and (2) the node numbering is incorrect forcing the elements to be pulled inside out. We must insure that $J > 0$, since J has a physical meaning: it is the ratio of the differential volume of the master element to the differential volume of the finite element. If the nodes are numbered correctly to insure that nodes are not pulled "inside out" (e.g., see Fig. 7.2) and that the elements are convex, then $J > 0$.

7.5.1 Good and Bad Elements: Examples

Let us consider a two-dimensional linear element shown in Fig. 3.5. The shape functions are:

- $\hat{\phi}_1 = \frac{1}{4}(1 - \zeta_1)(1 - \zeta_2)$,
- $\hat{\phi}_2 = \frac{1}{4}(1 + \zeta_1)(1 - \zeta_2)$,
- $\hat{\phi}_3 = \frac{1}{4}(1 + \zeta_1)(1 + \zeta_2)$,
- $\hat{\phi}_4 = \frac{1}{4}(1 - \zeta_1)(1 + \zeta_2)$.

The mapping functions are:

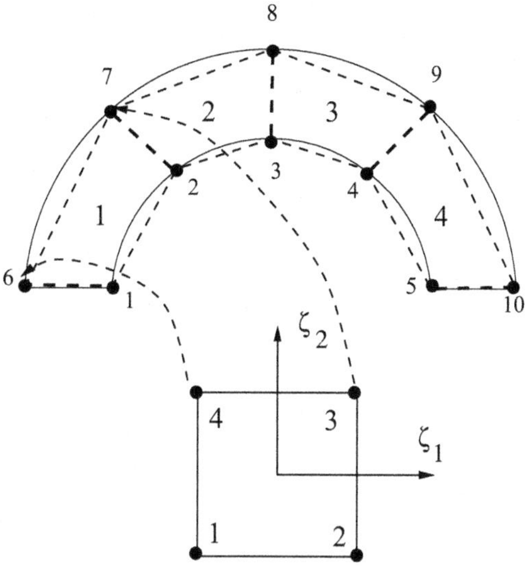

Fig. 7.2 An example of a mapped mesh for a semicircular strip

- $x_1 = \sum_{i=1}^{4} \mathcal{X}_{1i}\hat{\phi}_i \stackrel{\text{def}}{=} M_{x_1}(\zeta_1, \zeta_2)$,
- $x_2 = \sum_{i=1}^{4} \mathcal{X}_{2i}\hat{\phi}_i \stackrel{\text{def}}{=} M_{x_2}(\zeta_1, \zeta_2)$,

where $(\mathcal{X}_{1i}, \mathcal{X}_{2i})$ are true spatial coordinates of the ith node and where $\hat{\phi}(\zeta_1, \zeta_2) \stackrel{\text{def}}{=} \phi(x_1(\zeta_1, \zeta_2), x_2(\zeta_1, \zeta_2))$. Let us consider four examples. For the elements to be acceptable, the Jacobian corresponding to $\boldsymbol{F}$

$$d\boldsymbol{x} = \boldsymbol{F} \cdot d\boldsymbol{\zeta} \tag{7.16}$$

must be positive and finite throughout the element, where

$$J \stackrel{\text{def}}{=} |\boldsymbol{F}| \stackrel{\text{def}}{=} |\frac{\partial \boldsymbol{x}(x_1, x_2)}{\partial \boldsymbol{\zeta}(\zeta_1, \zeta_2)}| \quad \text{where} \quad \boldsymbol{F} \stackrel{\text{def}}{=} \begin{bmatrix} \frac{\partial x_1}{\partial \zeta_1} & \frac{\partial x_1}{\partial \zeta_2} \\ \frac{\partial x_2}{\partial \zeta_1} & \frac{\partial x_2}{\partial \zeta_2} \end{bmatrix}. \tag{7.17}$$

Explicitly,

$$J = |\boldsymbol{F}| = \frac{\partial x_1}{\partial \zeta_1}\frac{\partial x_2}{\partial \zeta_2} - \frac{\partial x_2}{\partial \zeta_1}\frac{\partial x_1}{\partial \zeta_2}. \tag{7.18}$$

For the four cases (Fig. 7.3), we have:

- Case 1: This element is acceptable, since $0 < J(\zeta_1, \zeta_2) < \infty$ throughout the element. The Jacobian is constant in this case.
- Case 2: This element is unacceptable, since $0 > J(\zeta_1, \zeta_2)$ throughout the element. The essential problem is that the nodes are numbered incorrectly, turning the element "inside out."

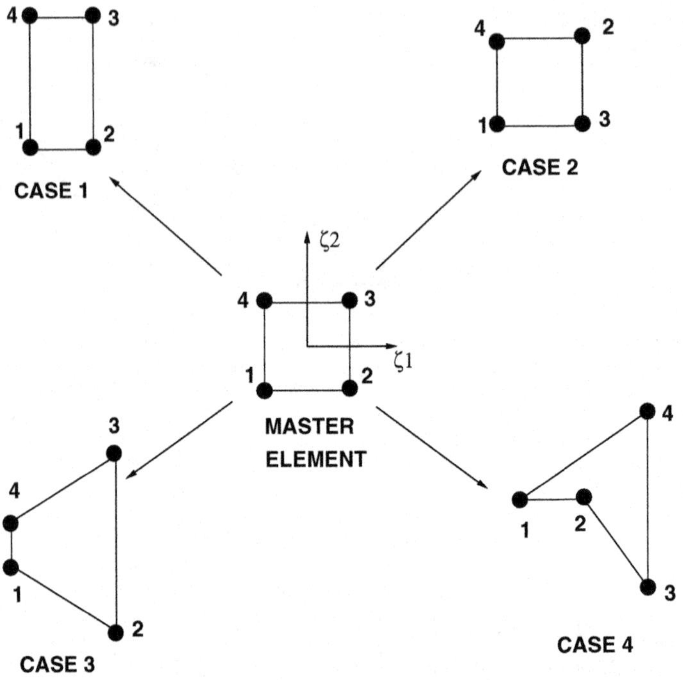

Fig. 7.3 A two-dimensional linear element and examples of mapping

- Case 3: This element is acceptable, since $0 < J(\zeta_1, \zeta_2) < \infty$ throughout the element. While the Jacobian is not constant throughout the element domain, it is positive and bounded.
- Case 4: This element is unacceptable, since $J(\zeta_1, \zeta_2) < 0$ in regions of the element. Even though the element is positive in some portions of the domain, a negative Jacobian in other parts can cause problems, such as potential singularities in the stiffness matrix.
- For linear elements, the key indicator for a problematic element is the nonconvexity of the element (even if numbered correctly).

7.6 Three-Dimensional Shape Functions

For the remainder of the monograph, we will illustrate the finite element method's construction with so-called trilinear "brick" elements. The master element shape functions form nodal bases of trilinear approximation given by:

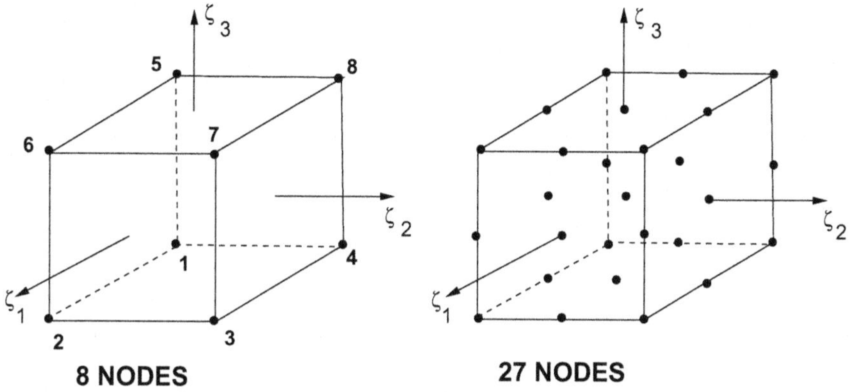

Fig. 7.4 Left: A trilinear eight-node hexahedron or "brick." Right: a 27-node element

$$\begin{array}{|ll|}\hline \hat{\phi}_1 = \frac{1}{8}(1-\zeta_1)(1-\zeta_2)(1-\zeta_3), & \hat{\phi}_2 = \frac{1}{8}(1+\zeta_1)(1-\zeta_2)(1-\zeta_3), \\ \hat{\phi}_3 = \frac{1}{8}(1+\zeta_1)(1+\zeta_2)(1-\zeta_3), & \hat{\phi}_4 = \frac{1}{8}(1-\zeta_1)(1+\zeta_2)(1-\zeta_3), \\ \hat{\phi}_5 = \frac{1}{8}(1-\zeta_1)(1-\zeta_2)(1+\zeta_3), & \hat{\phi}_6 = \frac{1}{8}(1+\zeta_1)(1-\zeta_2)(1+\zeta_3), \\ \hat{\phi}_7 = \frac{1}{8}(1+\zeta_1)(1+\zeta_2)(1+\zeta_3), & \hat{\phi}_8 = \frac{1}{8}(1-\zeta_1)(1+\zeta_2)(1+\zeta_3). \\ \hline \end{array} \tag{7.19}$$

For trilinear elements, we have a nodal basis consisting of eight nodes, and since it is vector valued, 24 total degrees of freedom (three degrees of freedom for each of the eight nodes).

Remark: For standard quadratic elements, we have a nodal basis consisting of 27 nodes (Fig. 7.4), and since it is vector valued, 81 total degrees of freedom (three degrees of freedom for each of the 27 nodes). The nodal shape functions can be derived quite easily, by realizing that it is a nodal basis, i.e., they are unity at the corresponding node, and zero at all other nodes, etc. For more details on the construction of higher-order elements, see Becker, Carey and Oden [1], Carey and Oden [2], Oden and Carey [3], Hughes [4], Bathe [5], and Zienkiewicz and Taylor [6].

7.7 Differential Properties of Shape Functions

We note that the ϕ_i's in the domain are never really computed. We actually start with the $\hat{\phi}_i$'s in the master domain. Therefore, in the stiffness matrix and right-hand side element calculations, all terms must be defined in terms of the local coordinates. With this in mind, we lay down some fundamental relations, which are directly related to the concepts of deformation presented in our discussion in continuum mechanics. It is not surprising that a deformation gradient reappears in the following form:

$$|\boldsymbol{F}| \stackrel{\text{def}}{=} |\frac{\partial \boldsymbol{x}(x_1, x_2, x_3)}{\partial \boldsymbol{\zeta}(\zeta_1, \zeta_2, \zeta_3)}| \quad \text{where} \quad \boldsymbol{F} \stackrel{\text{def}}{=} \begin{bmatrix} \frac{\partial x_1}{\partial \zeta_1} & \frac{\partial x_1}{\partial \zeta_2} & \frac{\partial x_1}{\partial \zeta_3} \\ \frac{\partial x_2}{\partial \zeta_1} & \frac{\partial x_2}{\partial \zeta_2} & \frac{\partial x_2}{\partial \zeta_3} \\ \frac{\partial x_3}{\partial \zeta_1} & \frac{\partial x_3}{\partial \zeta_2} & \frac{\partial x_3}{\partial \zeta_3} \end{bmatrix}, \tag{7.20}$$

where explicitly

$$|\boldsymbol{F}| = \frac{\partial x_1}{\partial \zeta_1}(\frac{\partial x_2}{\partial \zeta_2}\frac{\partial x_3}{\partial \zeta_3} - \frac{\partial x_3}{\partial \zeta_2}\frac{\partial x_2}{\partial \zeta_3}) - \frac{\partial x_1}{\partial \zeta_2}(\frac{\partial x_2}{\partial \zeta_1}\frac{\partial x_3}{\partial \zeta_3} - \frac{\partial x_3}{\partial \zeta_1}\frac{\partial x_2}{\partial \zeta_3}) + \frac{\partial x_1}{\partial \zeta_3}(\frac{\partial x_2}{\partial \zeta_1}\frac{\partial x_3}{\partial \zeta_2} - \frac{\partial x_3}{\partial \zeta_1}\frac{\partial x_2}{\partial \zeta_2}). \tag{7.21}$$

The differential relations $\boldsymbol{\zeta} \rightarrow \boldsymbol{x}$ are

$$\boxed{\begin{aligned} \frac{\partial}{\partial \zeta_1} &= \frac{\partial}{\partial x_1}\frac{\partial x_1}{\partial \zeta_1} + \frac{\partial}{\partial x_2}\frac{\partial x_2}{\partial \zeta_1} + \frac{\partial}{\partial x_3}\frac{\partial x_3}{\partial \zeta_1}, \\ \frac{\partial}{\partial \zeta_2} &= \frac{\partial}{\partial x_1}\frac{\partial x_1}{\partial \zeta_2} + \frac{\partial}{\partial x_2}\frac{\partial x_2}{\partial \zeta_2} + \frac{\partial}{\partial x_3}\frac{\partial x_3}{\partial \zeta_2}, \\ \frac{\partial}{\partial \zeta_3} &= \frac{\partial}{\partial x_1}\frac{\partial x_1}{\partial \zeta_3} + \frac{\partial}{\partial x_2}\frac{\partial x_2}{\partial \zeta_3} + \frac{\partial}{\partial x_3}\frac{\partial x_3}{\partial \zeta_3}. \end{aligned}} \tag{7.22}$$

The inverse differential relations $\boldsymbol{x} \rightarrow \boldsymbol{\zeta}$ are

$$\boxed{\begin{aligned} \frac{\partial}{\partial x_1} &= \frac{\partial}{\partial \zeta_1}\frac{\partial \zeta_1}{\partial x_1} + \frac{\partial}{\partial \zeta_2}\frac{\partial \zeta_2}{\partial x_1} + \frac{\partial}{\partial \zeta_3}\frac{\partial \zeta_3}{\partial x_1}, \\ \frac{\partial}{\partial x_2} &= \frac{\partial}{\partial \zeta_1}\frac{\partial \zeta_1}{\partial x_2} + \frac{\partial}{\partial \zeta_2}\frac{\partial \zeta_2}{\partial x_2} + \frac{\partial}{\partial \zeta_3}\frac{\partial \zeta_3}{\partial x_2}, \\ \frac{\partial}{\partial x_3} &= \frac{\partial}{\partial \zeta_1}\frac{\partial \zeta_1}{\partial x_3} + \frac{\partial}{\partial \zeta_2}\frac{\partial \zeta_2}{\partial x_3} + \frac{\partial}{\partial \zeta_3}\frac{\partial \zeta_3}{\partial x_3}, \end{aligned}} \tag{7.23}$$

and thus

$$\begin{Bmatrix} dx_1 \\ dx_2 \\ dx_3 \end{Bmatrix} = \underbrace{\begin{bmatrix} \frac{\partial x_1}{\partial \zeta_1} & \frac{\partial x_1}{\partial \zeta_2} & \frac{\partial x_1}{\partial \zeta_3} \\ \frac{\partial x_2}{\partial \zeta_1} & \frac{\partial x_2}{\partial \zeta_2} & \frac{\partial x_2}{\partial \zeta_3} \\ \frac{\partial x_3}{\partial \zeta_1} & \frac{\partial x_3}{\partial \zeta_2} & \frac{\partial x_3}{\partial \zeta_3} \end{bmatrix}}_{\boldsymbol{F}} \begin{Bmatrix} d\zeta_1 \\ d\zeta_2 \\ d\zeta_3 \end{Bmatrix} \tag{7.24}$$

and the inverse form

$$\begin{Bmatrix} d\zeta_1 \\ d\zeta_2 \\ d\zeta_3 \end{Bmatrix} = \underbrace{\begin{bmatrix} \frac{\partial \zeta_1}{\partial x_1} & \frac{\partial \zeta_1}{\partial x_2} & \frac{\partial \zeta_1}{\partial x_3} \\ \frac{\partial \zeta_2}{\partial x_1} & \frac{\partial \zeta_2}{\partial x_2} & \frac{\partial \zeta_2}{\partial x_3} \\ \frac{\partial \zeta_3}{\partial x_1} & \frac{\partial \zeta_3}{\partial x_2} & \frac{\partial \zeta_3}{\partial x_3} \end{bmatrix}}_{\boldsymbol{F}^{-1}} \begin{Bmatrix} dx_1 \\ dx_2 \\ dx_3 \end{Bmatrix}. \tag{7.25}$$

Noting the following relationship, from basic linear algebra

$$F^{-1}=\frac{adjF}{|F|}\quad \text{where}\quad adjF \stackrel{\text{def}}{=} \begin{bmatrix} A_{11} & A_{12} & A_{13} \\ A_{21} & A_{22} & A_{23} \\ A_{31} & A_{32} & A_{33} \end{bmatrix}^T, \tag{7.26}$$

where

$$\begin{array}{|ll|}\hline A_{11}=[\frac{\partial x_2}{\partial \zeta_2}\frac{\partial x_3}{\partial \zeta_3}-\frac{\partial x_3}{\partial \zeta_2}\frac{\partial x_2}{\partial \zeta_3}]=|F|\frac{\partial \zeta_1}{\partial x_1}, & A_{12}=-[\frac{\partial x_2}{\partial \zeta_1}\frac{\partial x_3}{\partial \zeta_3}-\frac{\partial x_3}{\partial \zeta_1}\frac{\partial x_2}{\partial \zeta_3}]=|F|\frac{\partial \zeta_2}{\partial x_1}, \\ A_{13}=[\frac{\partial x_2}{\partial \zeta_1}\frac{\partial x_3}{\partial \zeta_2}-\frac{\partial x_3}{\partial \zeta_1}\frac{\partial x_2}{\partial \zeta_2}]=|F|\frac{\partial \zeta_3}{\partial x_1}, & A_{21}=-[\frac{\partial x_1}{\partial \zeta_2}\frac{\partial x_3}{\partial \zeta_3}-\frac{\partial x_3}{\partial \zeta_2}\frac{\partial x_1}{\partial \zeta_3}]=|F|\frac{\partial \zeta_1}{\partial x_2}, \\ A_{22}=[\frac{\partial x_1}{\partial \zeta_1}\frac{\partial x_3}{\partial \zeta_3}-\frac{\partial x_3}{\partial \zeta_1}\frac{\partial x_1}{\partial \zeta_3}]=|F|\frac{\partial \zeta_2}{\partial x_2}, & A_{23}=-[\frac{\partial x_1}{\partial \zeta_1}\frac{\partial x_3}{\partial \zeta_2}-\frac{\partial x_3}{\partial \zeta_1}\frac{\partial x_1}{\partial \zeta_2}]=|F|\frac{\partial \zeta_3}{\partial x_2}, \\ A_{31}=[\frac{\partial x_1}{\partial \zeta_2}\frac{\partial x_2}{\partial \zeta_3}-\frac{\partial x_2}{\partial \zeta_2}\frac{\partial x_1}{\partial \zeta_3}]=|F|\frac{\partial \zeta_1}{\partial x_3}, & A_{32}=-[\frac{\partial x_1}{\partial \zeta_1}\frac{\partial x_2}{\partial \zeta_3}-\frac{\partial x_2}{\partial \zeta_1}\frac{\partial x_1}{\partial \zeta_3}]=|F|\frac{\partial \zeta_2}{\partial x_3}, \\ A_{33}=[\frac{\partial x_1}{\partial \zeta_1}\frac{\partial x_2}{\partial \zeta_2}-\frac{\partial x_2}{\partial \zeta_1}\frac{\partial x_1}{\partial \zeta_2}]=|F|\frac{\partial \zeta_3}{\partial x_3}. & \\ \hline\end{array} \tag{7.27}$$

With these relations, one can then solve for the components of F and F^{-1}.

7.8 Differentiation in the Referential Coordinates

We now need to express $[D]$ in terms $\zeta_1, \zeta_2, \zeta_3$, via

$$[D(\phi(x_1,x_2,x_3))]=[\hat{D}\left(\hat{\phi}(M_{x_1}(\zeta_1,\zeta_2,\zeta_3),M_{x_2}(\zeta_1,\zeta_2,\zeta_3),M_{x_3}(\zeta_1,\zeta_2,\zeta_3))\right)]. \tag{7.28}$$

Therefore, we write for the first column[2] of $[\hat{D}]$

$$\begin{bmatrix} \frac{\partial}{\partial \zeta_1}\frac{\partial \zeta_1}{\partial x_1}+\frac{\partial}{\partial \zeta_2}\frac{\partial \zeta_2}{\partial x_1}+\frac{\partial}{\partial \zeta_3}\frac{\partial \zeta_3}{\partial x_1} \\ 0 \\ 0 \\ \frac{\partial}{\partial \zeta_1}\frac{\partial \zeta_1}{\partial x_2}+\frac{\partial}{\partial \zeta_2}\frac{\partial \zeta_2}{\partial x_2}+\frac{\partial}{\partial \zeta_3}\frac{\partial \zeta_3}{\partial x_2} \\ 0 \\ \frac{\partial}{\partial \zeta_1}\frac{\partial \zeta_1}{\partial x_3}+\frac{\partial}{\partial \zeta_2}\frac{\partial \zeta_2}{\partial x_3}+\frac{\partial}{\partial \zeta_3}\frac{\partial \zeta_3}{\partial x_3} \end{bmatrix}, \tag{7.29}$$

[2]This is for illustration purposes only. For computational efficiency, one should not program such operations in this way. Clearly, the needless multiplication of zeros is to be avoided.

for the second column

$$\begin{bmatrix} 0 \\ \frac{\partial}{\partial \zeta_1}\frac{\partial \zeta_1}{\partial x_2} + \frac{\partial}{\partial \zeta_2}\frac{\partial \zeta_2}{\partial x_2} + \frac{\partial}{\partial \zeta_3}\frac{\partial \zeta_3}{\partial x_2} \\ 0 \\ \frac{\partial}{\partial \zeta_1}\frac{\partial \zeta_1}{\partial x_1} + \frac{\partial}{\partial \zeta_2}\frac{\partial \zeta_2}{\partial x_1} + \frac{\partial}{\partial \zeta_3}\frac{\partial \zeta_3}{\partial x_1} \\ \frac{\partial}{\partial \zeta_1}\frac{\partial \zeta_1}{\partial x_3} + \frac{\partial}{\partial \zeta_2}\frac{\partial \zeta_2}{\partial x_3} + \frac{\partial}{\partial \zeta_3}\frac{\partial \zeta_3}{\partial x_3} \\ 0 \end{bmatrix}, \tag{7.30}$$

and for the last column

$$\begin{bmatrix} 0 \\ 0 \\ \frac{\partial}{\partial \zeta_1}\frac{\partial \zeta_1}{\partial x_3} + \frac{\partial}{\partial \zeta_2}\frac{\partial \zeta_2}{\partial x_3} + \frac{\partial}{\partial \zeta_3}\frac{\partial \zeta_3}{\partial x_3} \\ 0 \\ \frac{\partial}{\partial \zeta_1}\frac{\partial \zeta_1}{\partial x_2} + \frac{\partial}{\partial \zeta_2}\frac{\partial \zeta_2}{\partial x_2} + \frac{\partial}{\partial \zeta_3}\frac{\partial \zeta_3}{\partial x_2} \\ \frac{\partial}{\partial \zeta_1}\frac{\partial \zeta_1}{\partial x_1} + \frac{\partial}{\partial \zeta_2}\frac{\partial \zeta_2}{\partial x_1} + \frac{\partial}{\partial \zeta_3}\frac{\partial \zeta_3}{\partial x_1} \end{bmatrix}. \tag{7.31}$$

For an element, our shape function matrix ($\stackrel{\text{def}}{=} [\hat{\phi}]$) has the following form for linear shape functions, for the first eight columns

$$\begin{bmatrix} \hat{\phi}_1 & \hat{\phi}_2 & \hat{\phi}_3 & \hat{\phi}_4 & \hat{\phi}_5 & \hat{\phi}_6 & \hat{\phi}_7 & \hat{\phi}_8 \\ 0 & 0 & 0 & 0 & 0 & 0 & 0 & 0 \\ 0 & 0 & 0 & 0 & 0 & 0 & 0 & 0 \end{bmatrix}, \tag{7.32}$$

for the second eight columns

$$\begin{bmatrix} 0 & 0 & 0 & 0 & 0 & 0 & 0 & 0 \\ \hat{\phi}_1 & \hat{\phi}_2 & \hat{\phi}_3 & \hat{\phi}_4 & \hat{\phi}_5 & \hat{\phi}_6 & \hat{\phi}_7 & \hat{\phi}_8 \\ 0 & 0 & 0 & 0 & 0 & 0 & 0 & 0 \end{bmatrix}, \tag{7.33}$$

and for the last eight columns

$$\begin{bmatrix} 0 & 0 & 0 & 0 & 0 & 0 & 0 & 0 \\ 0 & 0 & 0 & 0 & 0 & 0 & 0 & 0 \\ \hat{\phi}_1 & \hat{\phi}_2 & \hat{\phi}_3 & \hat{\phi}_4 & \hat{\phi}_5 & \hat{\phi}_6 & \hat{\phi}_7 & \hat{\phi}_8 \end{bmatrix}, \tag{7.34}$$

which in total is a 3×24 matrix. Therefore the product $[\hat{\boldsymbol{D}}][\hat{\phi}]$ is a 6×24 matrix of the form, for the first eight columns

$$\begin{bmatrix} \frac{\partial \hat{\phi}_1}{\partial \zeta_1}\frac{\partial \zeta_1}{\partial x_1} + \frac{\partial \hat{\phi}_1}{\partial \zeta_2}\frac{\partial \zeta_2}{\partial x_1} + \frac{\partial \hat{\phi}_1}{\partial \zeta_3}\frac{\partial \zeta_3}{\partial x_1}, \ldots 8 \\ 0\ 0\ 0\ 0\ 0\ 0\ 0\ 0 \\ 0\ 0\ 0\ 0\ 0\ 0\ 0\ 0 \\ \frac{\partial \hat{\phi}_1}{\partial \zeta_1}\frac{\partial \zeta_1}{\partial x_2} + \frac{\partial \hat{\phi}_1}{\partial \zeta_2}\frac{\partial \zeta_2}{\partial x_2} + \frac{\partial \hat{\phi}_1}{\partial \zeta_3}\frac{\partial \zeta_3}{\partial x_2}, \ldots 8 \\ 0\ 0\ 0\ 0\ 0\ 0\ 0\ 0 \\ \frac{\partial \hat{\phi}_1}{\partial \zeta_1}\frac{\partial \zeta_1}{\partial x_3} + \frac{\partial \hat{\phi}_1}{\partial \zeta_2}\frac{\partial \zeta_2}{\partial x_3} + \frac{\partial \hat{\phi}_1}{\partial \zeta_3}\frac{\partial \zeta_3}{\partial x_3}, \ldots 8 \end{bmatrix}, \tag{7.35}$$

for the second eight columns

$$\begin{bmatrix} 0\ 0\ 0\ 0\ 0\ 0\ 0\ 0 \\ \frac{\partial \hat{\phi}_1}{\partial \zeta_1}\frac{\partial \zeta_1}{\partial x_2} + \frac{\partial \hat{\phi}_1}{\partial \zeta_2}\frac{\partial \zeta_2}{\partial x_2} + \frac{\partial \hat{\phi}_1}{\partial \zeta_3}\frac{\partial \zeta_3}{\partial x_2}, \ldots 8 \\ 0\ 0\ 0\ 0\ 0\ 0\ 0\ 0 \\ \frac{\partial \hat{\phi}_1}{\partial \zeta_1}\frac{\partial \zeta_1}{\partial x_1} + \frac{\partial \hat{\phi}_1}{\partial \zeta_2}\frac{\partial \zeta_2}{\partial x_1} + \frac{\partial \hat{\phi}_1}{\partial \zeta_3}\frac{\partial \zeta_3}{\partial x_1}, \ldots 8 \\ \frac{\partial \hat{\phi}_1}{\partial \zeta_1}\frac{\partial \zeta_1}{\partial x_3} + \frac{\partial \hat{\phi}_1}{\partial \zeta_2}\frac{\partial \zeta_2}{\partial x_3} + \frac{\partial \hat{\phi}_1}{\partial \zeta_3}\frac{\partial \zeta_3}{\partial x_3}, \ldots 8 \\ 0\ 0\ 0\ 0\ 0\ 0\ 0\ 0\ , \end{bmatrix}, \tag{7.36}$$

and for the last eight columns

$$\begin{bmatrix} 0\ 0\ 0\ 0\ 0\ 0\ 0\ 0 \\ 0\ 0\ 0\ 0\ 0\ 0\ 0\ 0 \\ \frac{\partial \hat{\phi}_1}{\partial \zeta_1}\frac{\partial \zeta_1}{\partial x_3} + \frac{\partial \hat{\phi}_1}{\partial \zeta_2}\frac{\partial \zeta_2}{\partial x_3} + \frac{\partial \hat{\phi}_1}{\partial \zeta_3}\frac{\partial \zeta_3}{\partial x_3}, \ldots 8 \\ 0\ 0\ 0\ 0\ 0\ 0\ 0\ 0 \\ \frac{\partial \hat{\phi}_1}{\partial \zeta_1}\frac{\partial \zeta_1}{\partial x_2} + \frac{\partial \hat{\phi}_1}{\partial \zeta_2}\frac{\partial \zeta_2}{\partial x_2} + \frac{\partial \hat{\phi}_1}{\partial \zeta_3}\frac{\partial \zeta_3}{\partial x_2}, \ldots 8 \\ \frac{\partial \hat{\phi}_1}{\partial \zeta_1}\frac{\partial \zeta_1}{\partial x_1} + \frac{\partial \hat{\phi}_1}{\partial \zeta_2}\frac{\partial \zeta_2}{\partial x_1} + \frac{\partial \hat{\phi}_1}{\partial \zeta_3}\frac{\partial \zeta_3}{\partial x_1}, \ldots 8. \end{bmatrix}. \tag{7.37}$$

Finally, with quadrature for each element, we can form each of the element contributions for $[\boldsymbol{K}]\{\boldsymbol{a}\} = \{\boldsymbol{R}\}$:

- For the stiffness matrix:

$$[K^e] = \underbrace{\sum_{q=1}^{g}\sum_{r=1}^{g}\sum_{s=1}^{g} w_q w_r w_s ([\hat{\boldsymbol{D}}][\hat{\phi}])^T [\hat{\mathbb{E}}]([\hat{\boldsymbol{D}}][\hat{\phi}])|\boldsymbol{F}|}_{\text{standard}}$$

$$+ \underbrace{\sum_{q=1}^{g}\sum_{r=1}^{g} w_q w_r P^{\star} [\hat{\phi}]^T [\hat{\phi}] |\boldsymbol{F}_s|}_{\text{penalty for } \Gamma_u \cap \partial\Omega_e \neq 0}, \tag{7.38}$$

- For the load vector:

$$\{R^e\} = \underbrace{\sum_{q=1}^{g}\sum_{r=1}^{g}\sum_{s=1}^{g} w_q w_r w_s [\hat{\phi}]^T \{f\} |\boldsymbol{F}|}_{\text{standard}}$$

$$+ \underbrace{\sum_{q=1}^{g}\sum_{r=1}^{g} w_q w_r [\hat{\phi}]^T \{\boldsymbol{t}^*\} |\boldsymbol{F}_s|}_{\text{for } \Gamma_t \cap \partial\Omega_e \neq 0} + \underbrace{\sum_{q=1}^{g}\sum_{r=1}^{g} w_q w_r P^{\star} [\hat{\phi}]^T \{\boldsymbol{u}^*\} |\boldsymbol{F}_s|}_{\text{penalty for } \Gamma_u \cap \partial\Omega_e \neq 0}, \tag{7.39}$$

where w_q, etc., are Gauss weights and where $|\boldsymbol{F}_s|$ represents the (surface) Jacobians of element faces on the exterior surface of the body, where, depending on the surface on which it is to be evaluated upon, one of the ζ components will be $+1$ or -1. These surface Jacobians can be evaluated in a variety of ways, for example using Nanson's formula, which is derived in Appendix B and which is discussed further shortly.

7.8.1 Implementation Issues

Following similar procedures as for one-dimensional problems, the global stiffness matrix $\boldsymbol{K}(I, J)$ can be efficiently stored in an element-by-element manner via $\boldsymbol{k}(e, i, j)$, i and j are the local entries in element number e. The amount of memory required with this relatively simple storage system is, for trilinear hexahedra, $\boldsymbol{k}(e, 24, 24) = 576$ times the number of finite elements, where the $\boldsymbol{k}$ are the individual element stiffness matrices. If matrix symmetry is taken into account, the memory requirements are 300 times the number of finite elements. As in one-dimension, this simple approach is so-called element-by-element storage. The element-by-element storage is critical in this regard to reduce the memory requirements.[3] For an element-by-element storage scheme, a global/local index relation must be made to connect the

[3]If a direct storage of the finite element storage of the stiffness matrix were attempted, the memory requirements would be $\boldsymbol{K}(DOF, DOF) = DOF \times DOF$, where DOF indicates the total degrees of freedom, which for large problems, would be extremely demanding.

local entry to the global entry for the subsequent linear algebraic solution processes. This is a relatively simple and efficient storage system to encode. The element-by-element strategy has other advantages with regard to element-by-element system CG solvers, as introduced earlier. The actual computation cost of the matrix-vector multiplication in an element-by-element CG method is a $[24 \times 24]$ matrix times a $\{24 \times 1\}$ vector times the number of elements. This is an $\mathcal{O}(N)$ calculation. If we consider $\mathcal{I}$ iterations necessary for convergence below an error tolerance, then the entire operation costs are $\mathcal{O}(\mathcal{I}N)$.

7.8.2 An Example of the Storage Scaling

Element-by-element storage has reduced the storage requirements dramatically. For example, consider a cube meshed uniformly with M elements in each direction (Fig. 7.5), thus $(M + 1)^3$ nodes and $3(M + 1)^3$ degrees of freedom for elasticity problems. A comparison of storage yields:

- Direct storage: $3(M + 1)^3 \times 3(M + 1)^3 = 9\mathcal{O}(M^6)$,
- Element-by-element storage: $M^3 \times 24 \times 24 = 576M^3$, and
- Element-by-element storage with symmetry reduction: $300M^3$.

Clearly, a ratio of direct storage to element-by-element storage scales as cubically $\mathcal{O}(M^3)$. Thus,

- Direct/element-by-element storage ratio $\approx \frac{9\mathcal{O}(M^6)}{300M^3} = \frac{3}{100}\mathcal{O}(M^3)$ and

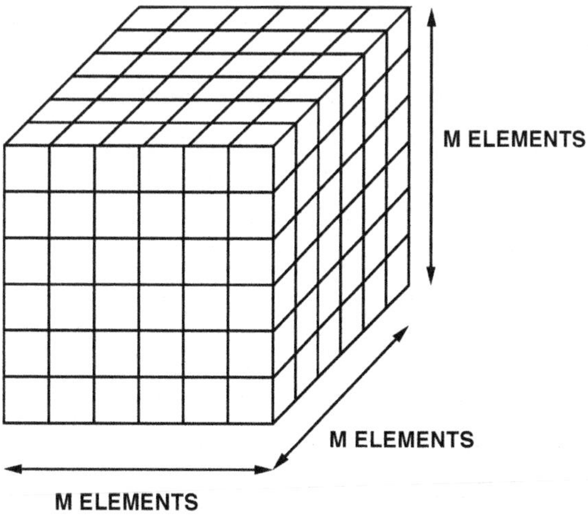

Fig. 7.5 A cube with M elements in each directions

- Direct/element-by-element solving ratio $\approx \frac{\mathcal{O}(N^3)}{300\mathcal{I}\mathcal{O}(N)} = \frac{1}{300\mathcal{I}}\mathcal{O}(N^2) = \frac{1}{300\mathcal{I}}\mathcal{O}((3(M+1)^3)^2) = \frac{9}{300\mathcal{I}}\mathcal{O}((M+1)^6)$,
- For $M = 10^2$, direct/element-by-element storage ratio $\approx 3 \times 10^4$ and
- For $M = 10^2$, direct/element-by-element solving ratio $\approx \frac{9}{300\mathcal{I}}\mathcal{O}(10^{12})$.

Of course there are other compact storage schemes, and we refer the reader to the references for details.

7.9 Surface Jacobians and Nanson's Formula

In order to compute surface integrals, for a general element that intersects the exterior surface, one must (Fig. 7.6):

1. Identify which element face of the master element corresponds to that surface. One of the ζ-coordinates must be set to ± 1; i.e., ζ_1, ζ_2, and ζ_3 must be set equal to ± 1 for the faces that correspond to the exposed surfaces on the body where boundary conditions are imposed. Generally, we seek to integrate a quantity, $\boldsymbol{Q}$, over the surface of the actual, deformed, element by computing over the master element, for which we can use standard Gaussian quadrature:

$$\int_{\partial\Omega_e} \boldsymbol{Q}\, dA_e = \int_{\partial\hat{\Omega}_e} \hat{\boldsymbol{Q}}\, d\hat{A}_e, \tag{7.40}$$

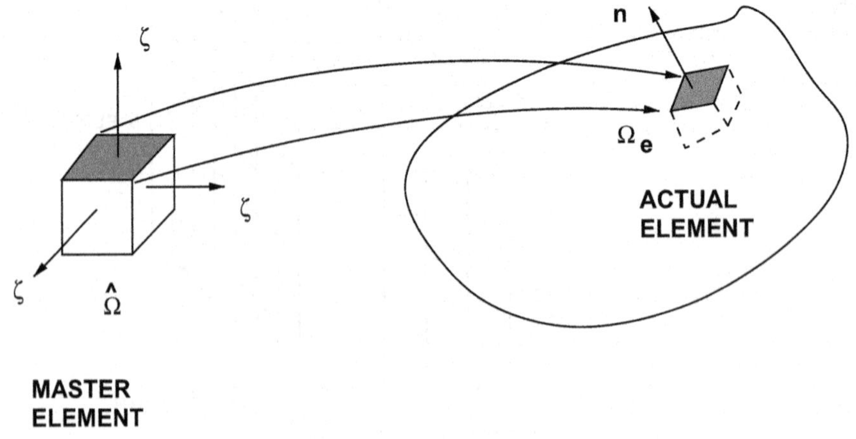

Fig. 7.6 Use of Nanson's formula for surface integration

2. Using Nanson's formula, $\boldsymbol{n}dA_e = \boldsymbol{J}\boldsymbol{F}^{-T}\cdot\boldsymbol{N}d\hat{A}_e$; thus, $dA_e = (\boldsymbol{J}\boldsymbol{F}^{-T}\cdot\boldsymbol{N})\cdot\boldsymbol{n}d\hat{A}_e = J_s d\hat{A}_e$, where $d\hat{A}_e = d\zeta_i d\zeta_j$ is the differential element area on a master element.
3. Identify the normal at a Gauss point on the surface, and ensure that one of the ζ coordinates is set to ± 1.

7.10 Post-Processing

Post-processing for the stress, strain, and energy from the existing displacement solution, i.e., the values of the nodal displacements, the shape functions, is straightforward. Essentially the process is the same as the formation of the system to be solved. Therefore, for each element

$$\left\{\begin{array}{c}\epsilon_{11}^h\\ \epsilon_{22}^h\\ \epsilon_{33}^h\\ 2\epsilon_{12}^h\\ 2\epsilon_{23}^h\\ 2\epsilon_{13}^h\end{array}\right\} = \left[\begin{array}{ccc}\frac{\partial}{\partial x_1} & 0 & 0\\ 0 & \frac{\partial}{\partial x_2} & 0\\ 0 & 0 & \frac{\partial}{\partial x_3}\\ \frac{\partial}{\partial x_2} & \frac{\partial}{\partial x_1} & 0\\ 0 & \frac{\partial}{\partial x_3} & \frac{\partial}{\partial x_2}\\ \frac{\partial}{\partial x_3} & 0 & \frac{\partial}{\partial x_1}\end{array}\right] \underbrace{\left\{\begin{array}{c}\sum_{i=1}^8 a_{1i}\phi_i\\ \sum_{i=1}^8 a_{2i}\phi_i\\ \sum_{i=1}^8 a_{3i}\phi_i\end{array}\right\}}_{\text{known values}} \tag{7.41}$$

where the a_{1i}, a_{2i}, and a_{3i} are the values at the node i for the x_1, x_2, and x_3 components, and *where the global coordinates must be transformed to the master system, in both the deformation tensor and the displacement representation.* Typically, within each element, at each Gauss point, we add up all eight contributions (from the basis functions) for each of the six components and then multiply by the corresponding nodal displacements that have previously been calculated. Gauss point locations are the preferred location to post-process the solution since they typically exhibit so-called superconvergent properties (more accurate than the theoretical estimates). In other words, they are usually the most accurate locations of the finite element approximation (see Ainsworth and Oden [7], Zienkiewicz and Taylor [6], and Zienkiewicz and Zhu [8]). The following expressions must be evaluated at the Gauss points, multiplied by the appropriate weights and added together:

$$\boxed{\begin{array}{lll}\frac{\partial u_1^h}{\partial x_1} = \sum_{i=1}^8 a_{1i}\frac{\partial\phi_i}{\partial x_1}, & \frac{\partial u_2^h}{\partial x_1} = \sum_{i=1}^8 a_{2i}\frac{\partial\phi_i}{\partial x_1}, & \frac{\partial u_3^h}{\partial x_1} = \sum_{i=1}^8 a_{3i}\frac{\partial\phi_i}{\partial x_1},\\ \frac{\partial u_1^h}{\partial x_2} = \sum_{i=1}^8 a_{1i}\frac{\partial\phi_i}{\partial x_2}, & \frac{\partial u_2^h}{\partial x_2} = \sum_{i=1}^8 a_{2i}\frac{\partial\phi_i}{\partial x_2}, & \frac{\partial u_3^h}{\partial x_2} = \sum_{i=1}^8 a_{3i}\frac{\partial\phi_i}{\partial x_2},\\ \frac{\partial u_1^h}{\partial x_3} = \sum_{i=1}^8 a_{1i}\frac{\partial\phi_i}{\partial x_3}, & \frac{\partial u_2^h}{\partial x_3} = \sum_{i=1}^8 a_{2i}\frac{\partial\phi_i}{\partial x_3}, & \frac{\partial u_3^h}{\partial x_3} = \sum_{i=1}^8 a_{3i}\frac{\partial\phi_i}{\partial x_3},\end{array}} \tag{7.42}$$

where a_{1i} denotes the x_1 component of the displacement of the ith node. Combining the numerical derivatives to form the strains we obtain $\epsilon^h_{11} = \frac{\partial u^h_1}{\partial x_1}$, $\epsilon^h_{22} = \frac{\partial u^h_2}{\partial x_2}$, $\epsilon^h_{33} = \frac{\partial u^h_3}{\partial x_3}$ and $2\epsilon^h_{12} = \gamma_{12} = \frac{\partial u^h_1}{\partial x_2} + \frac{\partial u^h_2}{\partial x_1}$, $2\epsilon^h_{23} = \gamma_{23} = \frac{\partial u^h_2}{\partial x_3} + \frac{\partial u^h_3}{\partial x_2}$, and $2\epsilon^h_{13} = \gamma_{13} = \frac{\partial u^h_1}{\partial x_3} + \frac{\partial u^h_3}{\partial x_1}$.

References

1. Becker, E. B., Carey, G. F., & Oden, J. T. (1980). *Finite elements: An introduction*. Englewood Cliffs: Prentice-Hall.
2. Carey, G. F., & Oden, J. T. (1983). *Finite elements: A second course*. Englewood Cliffs: Prentice-Hall.
3. Oden, J. T., & Carey, G. F. (1984). *Finite elements: Mathematical aspects*. Englewood Cliffs: Prentice-Hall.
4. Hughes, T. J. R. (1989). *The finite element method*. Englewood Cliffs: Prentice Hall.
5. Bathe, K. J. (1996). *Finite element procedures*. Englewood Cliffs: Prentice-Hall.
6. Zienkiewicz, O. C., & Taylor, R. L. (1991). *The finite element method* (Vol. I and II). New York: McGraw-Hill.
7. Ainsworth, M., & Oden, J. T. (2000). *A posterori error estimation in finite element analysis*. New York: Wiley.
8. Zienkiewicz, O. C., & Zhu, J. Z. (1987). A simple error estimator and adaptive procedure for practical engineering analysis. *International Journal for Numerical Methods in Engineering*, *24*, 337–357.

Accuracy of the Finite Element Method in Three Dimensions

8

8.1 Introduction

As we have seen in the one-dimensional analysis, the essential idea in the finite element method is to select a finite dimensional subspatial approximation of the true solution and form the following weak boundary problem:

$$\begin{array}{|l|}\hline \text{Find } \boldsymbol{u}^h \in \boldsymbol{H}^h_u(\Omega) \subset \boldsymbol{H}^1(\Omega), \text{ with } \boldsymbol{u}^h|_{\Gamma_u} = \boldsymbol{u}^*, \text{ such that} \\ \underbrace{\int_\Omega \nabla \boldsymbol{\nu}^h : \mathbb{E} : \nabla \boldsymbol{u}^h \, d\Omega}_{\mathcal{B}(\boldsymbol{u}^h, \boldsymbol{\nu}^h)} = \underbrace{\int_\Omega \boldsymbol{f} \cdot \boldsymbol{\nu}^h \, d\Omega + \int_{\Gamma_t} \boldsymbol{t}^* \cdot \boldsymbol{\nu}^h \, dA}_{\mathcal{F}(\boldsymbol{\nu}^h)}, \\ \forall \boldsymbol{\nu}^h \in \boldsymbol{H}^h_\upsilon(\Omega) \subset \boldsymbol{H}^1(\Omega), \text{ with } \boldsymbol{\nu}^h|_{\Gamma_u} = \boldsymbol{0}. \\ \hline \end{array} \tag{8.1}$$

The critical point is that $\boldsymbol{H}^h_u(\Omega), \boldsymbol{H}^h_\upsilon(\Omega) \subset \boldsymbol{H}^1(\Omega)$. This "inner" approximation allows the development of straightforward subspatial error estimates. We will choose $\boldsymbol{H}^h_u(\Omega)$ and $\boldsymbol{H}^h_\upsilon(\Omega)$ to coincide. We have for any kinematically admissible function, $\boldsymbol{w}$, a definition of the so-called energy norm

$$||\boldsymbol{u} - \boldsymbol{w}||^2_{E(\Omega)} \stackrel{\text{def}}{=} \int_\Omega (\nabla \boldsymbol{u} - \nabla \boldsymbol{w}) : \mathbb{E} : (\nabla \boldsymbol{u} - \nabla \boldsymbol{w}) \, d\Omega = \mathcal{B}(\boldsymbol{u} - \boldsymbol{w}, \boldsymbol{u} - \boldsymbol{w}). \tag{8.2}$$

Note that in the event that nonconstant displacements are specified on the boundary, then $\boldsymbol{u} - \boldsymbol{w} = constant$ is unobtainable unless $\boldsymbol{u} - \boldsymbol{w} = \boldsymbol{0}$, and the semi-norm in Eq. (8.2) is a norm in the strict mathematical sense. Under standard assumptions the fundamental a priori error estimate for the finite element method is

$$||\boldsymbol{u} - \boldsymbol{u}^h||_{E(\Omega)} \leq \mathcal{C}(\boldsymbol{u}, p) h^{min(r-1,p)} \stackrel{\text{def}}{=} \gamma, \tag{8.3}$$

T. I. Zohdi, *A Finite Element Primer for Beginners*, The Basics,
https://doi.org/10.1007/978-3-319-70428-9_8

where p is the (complete) polynomial order of the finite element method used, r is the regularity of the exact solution, and $\mathcal{C}$ is a global constant dependent on the exact solution and the polynomial approximation. $\mathcal{C}$ is independent of h, the maximum element diameter. For details see, Ainsworth and Oden [1], Becker, Carey and Oden [2], Carey and Oden [3], Oden and Carey [4], Hughes [5], Szabo and Babuska [6], and Bathe [7] for more mathematically precise treatments.

Remark: As we have mentioned previously, we note that the set of functions specified by $\boldsymbol{H}^h_u(\Omega) \subset \boldsymbol{H}^1(\Omega)$ with $\boldsymbol{u}^h|_{\Gamma_u} = \boldsymbol{u}^*$ is technically not a space of functions and should be characterized as "a linear variety." This does not pose a problem for the ensuing analysis; however, for precise mathematical details, see Oden and Demkowicz [8].

8.2 The "Best Approximation" Theorem

As in the one-dimensional case we have

$$\mathcal{B}(\boldsymbol{u}, \boldsymbol{\nu}) = \mathcal{F}(\boldsymbol{\nu}), \tag{8.4}$$

$\forall \boldsymbol{\nu} \in \boldsymbol{H}^1(\Omega)$ and

$$\mathcal{B}(\boldsymbol{u}^h, \boldsymbol{\nu}^h) = \mathcal{F}(\boldsymbol{\nu}^h), \tag{8.5}$$

$\forall \boldsymbol{\nu}^h \in \boldsymbol{H}^h_v(\Omega) \subset \boldsymbol{H}^1(\Omega)$. Subtracting Eq. 8.5 from 8.4 implies a Galerkin-like (Fig. 1.1) orthogonality property of "inner approximations"

$$\mathcal{B}(\boldsymbol{u} - \boldsymbol{u}^h, \boldsymbol{\nu}^h) = \mathcal{B}(\boldsymbol{e}^h, \boldsymbol{\nu}^h) = 0, \qquad \forall \boldsymbol{\nu}^h \in \boldsymbol{H}^h_v(\Omega) \subset \boldsymbol{H}^1(\Omega), \tag{8.6}$$

where the error is defined by $\boldsymbol{e}^h \stackrel{\text{def}}{=} \boldsymbol{u} - \boldsymbol{u}^h$. An important observation is that

$$\boldsymbol{e}^h - \boldsymbol{\nu}^h = \boldsymbol{u} - \boldsymbol{u}^h - \boldsymbol{\nu}^h = \boldsymbol{u} - \boldsymbol{z}^h, \tag{8.7}$$

thus

$$\mathcal{B}(\boldsymbol{e}^h - \boldsymbol{\nu}^h, \boldsymbol{e}^h - \boldsymbol{\nu}^h) = \mathcal{B}(\boldsymbol{e}^h, \boldsymbol{e}^h) - 2\mathcal{B}(\boldsymbol{e}^h, \boldsymbol{\nu}^h) + \underbrace{\mathcal{B}(\boldsymbol{\nu}^h, \boldsymbol{\nu}^h)}_{\geq 0}, \tag{8.8}$$

which implies

$$\mathcal{B}(\boldsymbol{u} - \boldsymbol{u}^h, \boldsymbol{u} - \boldsymbol{u}^h) \leq \mathcal{B}(\boldsymbol{u} - \boldsymbol{z}^h, \boldsymbol{u} - \boldsymbol{z}^h). \tag{8.9}$$

This implies that the FEM-constructed solution is the best possible in the energy norm (Fig. 8.1).

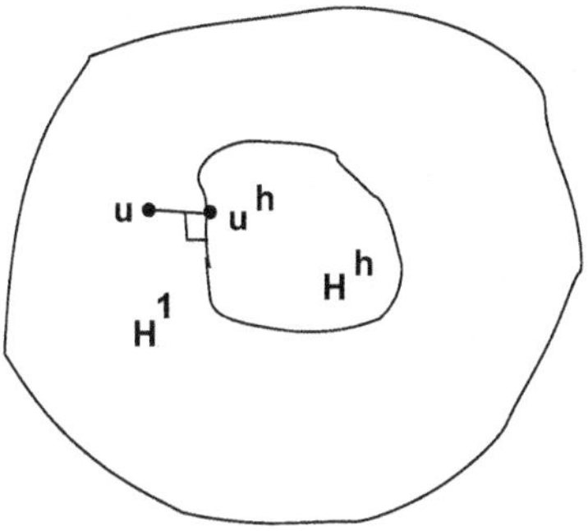

Fig. 8.1 An illustration of the best approximation theorem

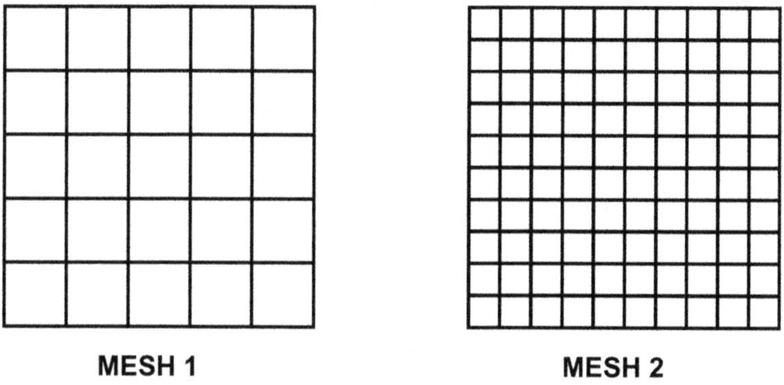

Fig. 8.2 Successively refined (halved/embedded) meshes used to estimate the error

8.3 Simple Estimates for Adequate FEM Meshes Revisited for Three Dimensions

As stated earlier, under standard assumptions the classical a priori error estimate for the finite element method is (Eq. 8.3), $||\boldsymbol{u} - \boldsymbol{u}^h||_{E(\Omega)} \leq C(\boldsymbol{u}, p)h^{min(r-1,p) \stackrel{\text{def}}{=} \gamma}$. Using the PMPE for a finite element solution (Eq. 6.14), with $\boldsymbol{w} = \boldsymbol{u}^h$, we have

$$||\boldsymbol{u} - \boldsymbol{u}^h||^2_{E(\Omega)} = 2(\mathcal{J}(\boldsymbol{u}^h) - \mathcal{J}(\boldsymbol{u})). \tag{8.10}$$

By solving the boundary value problem associated for two successively finer meshes, $h_1 > h_2$, with the following property $\mathcal{J}(\boldsymbol{u}^{h_1}) \geq \mathcal{J}(\boldsymbol{u}^{h_2}) \geq \mathcal{J}(\boldsymbol{u}^{h=0})$, we can set up the following system of equations for unknown constant C (Fig. 8.2):

$$\begin{aligned} ||\boldsymbol{u} - \boldsymbol{u}^{h_1}||^2_{E(\Omega)} &= 2(\mathcal{J}(\boldsymbol{u}^{h_1}) - \mathcal{J}(\boldsymbol{u}^h)) \approx C^2 h_1^{2\gamma}, \\ ||\boldsymbol{u} - \boldsymbol{u}^{h_2}||^2_{E(\Omega)} &= 2(\mathcal{J}(\boldsymbol{u}^{h_2}) - \mathcal{J}(\boldsymbol{u}^h)) \approx C^2 h_2^{2\gamma}. \end{aligned} \tag{8.11}$$

Solving for C

$$C = \sqrt{\frac{2(\mathcal{J}(\boldsymbol{u}^{h_1}) - \mathcal{J}(\boldsymbol{u}^{h_2}))}{h_1^{2\gamma} - h_2^{2\gamma}}}. \tag{8.12}$$

One can now solve for the appropriate mesh size by writing

$$Ch_{tol}^{\gamma} \approx TOL \Rightarrow h_{tol} \approx \left(\frac{TOL}{C}\right)^{\frac{1}{\gamma}}. \tag{8.13}$$

In summary, to monitor the discretization error, we apply the following (Fig. 8.2) algorithm ($K = 0.5$)

$$\begin{array}{|l|}\hline \textbf{STEP 1}: \text{SOLVE WITH COARSE MESH} = \text{h}_1 \Rightarrow \boldsymbol{u}^{h_1} \Rightarrow \mathcal{J}(\boldsymbol{u}^{h_1}) \\ \textbf{STEP 2}: \text{SOLVE WITH FINER MESH} = \text{h}_2 = \text{K} \times \text{h}_1 \Rightarrow \boldsymbol{u}^{h_2} \Rightarrow \mathcal{J}(\boldsymbol{u}^{h_2}) \\ \textbf{STEP 3}: \text{COMPUTE C} \Rightarrow h_{tol} \approx \left(\frac{TOL}{C}\right)^{\frac{1}{\gamma}}. \\ \hline \end{array} \tag{8.14}$$

Remarks: As for one-dimensional problems, while this scheme provides a simple estimate for the global mesh fineness needed, the meshes need to be locally refined to ensure tolerable accuracy throughout the domain.

8.4 Local Error Estimation and Adaptive Mesh Refinement

To drive local mesh refinement schemes there are a variety of error estimation procedures. We mention the two main ones: *recovery methods and residual methods*.

8.4.1 A Posteriori Recovery Methods

The so-called recovery methods are based on the assumption that there is a function $G(\boldsymbol{u}^h)$ that is closer to $\nabla \boldsymbol{u}$ than $\nabla \boldsymbol{u}^h$, which can be used to estimate the error. The most popular of these is the Zienkiewicz–Zhu [9] estimator. Zienkiewicz and Zhu developed an error estimation technique that is effective for a wide class of problems. It is based on the notion that gradients of the solution obtained on a given mesh can be smoothed and compared with the original solution to assess the error. The sampling points at which the gradient's error is to be evaluated are so-called superconvergent points where the convergence is above the theoretical optimum. However, these points must be searched for and may not even exist, i.e., superconvergence occurs only in very special situations. By superconvergence, we mean that the exponent is higher than the standard theoretical estimate (θ):

$$||\boldsymbol{u} - \boldsymbol{u}^h||_{H^s(\Omega)} \leq C(\boldsymbol{u}, p)h^{min(p+1-s,r-s)} \stackrel{\text{def}}{=} \theta \tag{8.15}$$

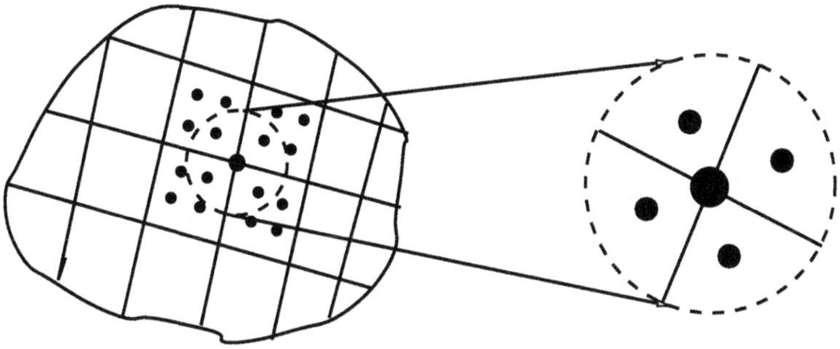

Fig. 8.3 The Zienkiewicz–Zhu error estimator takes the solution at neighboring Gauss points to estimate the error at a node

The function G is obtained by calculating a least squares fit to the gradient of a sample superconvergent points (potentially several hundred in three dimensions) of elements surrounding a finite element node (Fig. 8.3). The new gradient then serves to estimate the error locally over a "patch" of elements, i.e., a group of element sharing a common node,

$$||G(\boldsymbol{u}^h) - \nabla \boldsymbol{u}^h||_{patch} \approx error \tag{8.16}$$

This is by far the most popular method in the engineering community to estimate the error and also has the benefit of post-processing stresses as a by-product.

8.4.2 A Posteriori Residual Methods

Residual methods require no a posteriori system of equations to be solved. Such methods bound the error by making use of

- the FEM solution itself,
- the data on the boundary,
- the error equation, and
- the Galerkin orthogonality property.

As in the one-dimensional case discussed earlier, the approach is to form the following bound

$$||\boldsymbol{u} - \boldsymbol{u}^h||^2_{E(\Omega)} \leq C_1 \sum_{e=1}^{N} \underbrace{h_e^2 ||\boldsymbol{r}_1||^2_{L^2(\Omega_e)}}_{interior} + C_2 \sum_{I=1}^{INT} \underbrace{h_{eI} ||[\![\boldsymbol{r}_2]\!]||^2_{L^2(\partial\Omega_I)}}_{interfaces} + C_3 \sum_{J=1}^{B-INT} \underbrace{h_{eJ} ||\boldsymbol{r}_3||^2_{L^2(\partial\Omega_{JB})}}_{exterior-boundary}. \tag{8.17}$$

where

- C_1, C_2, and C_3 are constants,
- h_e are the sizes of the elements,
- the interior element residual is $\boldsymbol{r}_1 = \nabla \cdot \boldsymbol{\sigma}^h + \boldsymbol{f}$,
- the interior interface "jump" residual is $[\![\boldsymbol{r}_2]\!] = [\![\boldsymbol{t}]\!]$,
- the boundary interface ("dissatisfaction") residual is $\boldsymbol{r}_3 = \boldsymbol{\sigma}^h \cdot \boldsymbol{n} - \boldsymbol{t}^*$ and
- local error indicators are defined by

$$\zeta_e^2 \stackrel{\text{def}}{=} C_1 h_e^2 ||\boldsymbol{r}_1||^2_{L^2(\Omega_e)} + C_2 h_{eI} ||[\![\boldsymbol{r}_2]\!]||^2_{L^2(\partial\Omega_I)} + C_3 h_{eJ} ||\boldsymbol{r}_3||^2_{L^2(\partial\Omega_{JB})}. \quad (8.18)$$

The local quantities ζ_e are used to decide whether an element is to be refined (Fig. 4.3). If $\zeta_e > TOL$, then the element is refined. Such estimates, used to guide local adaptive finite element mesh refinement techniques, were first developed in Babúska and Rheinboldt [10] for one-dimensional problems and in Babùska and Miller [11] and Kelly et al. [12] for two-dimensional problems. For reviews see Ainsworth and Oden [1].

References

1. Ainsworth, M., & Oden, J. T. (2000). *A posterori error estimation in finite element analysis.* New York: Wiley.
2. Becker, E. B., Carey, G. F., & Oden, J. T. (1980). *Finite elements: An introduction.* Englewood Cliffs: Prentice Hall.
3. Carey, G. F., & Oden, J. T. (1983). *Finite elements: A second course.* Englewood Cliffs: Prentice Hall.
4. Oden, J. T., & Carey, G. F. (1984). *Finite elements: Mathematical aspects.* Englewood Cliffs: Prentice Hall.
5. Hughes, T. J. R. (1989). *The finite element method.* Englewood Cliffs: Prentice Hall.
6. Szabo, B., & Babúska, I. (1991). *Finite element analysis.* New York: Wiley Interscience.
7. Bathe, K. J. (1996). *Finite element procedures.* Englewood Cliffs: Prentice Hall.
8. Oden, J. T., & Demkowicz, L. F. (2010). *Applied functional analysis.* Boca Raton: CRC Press.
9. Zienkiewicz, O. C., & Zhu, J. Z. (1987). A simple error estimator and adaptive procedure for practical engineering analysis. *International Journal for Numerical Methods in Engineering, 24*, 337–357.
10. Babúska, I., & Rheinbolt, W. C. (1978). A posteriori error estimates for the finite element method. *The International Journal for Numerical Methods in Engineering, 12*, 1597–1615.
11. Babúska, I., & Miller, A. D. (1987). A feedback finite element method with a-posteriori error estimation. Part I. *Computer Methods in Applied Mechanics and Engineering, 61*, 1–40.
12. Kelly, D. W., Gago, J. R., Zienkiewicz, O. C., & Babùska, I. (1983). A posteriori error analysis and adaptive processes in the finite element method. Part I-error analysis. *International Journal for Numerical Methods in Engineering, 19*, 1593–1619.

9 Time-Dependent Problems

9.1 Introduction

We now give a brief introduction to time-dependent problems through the equations of elastodynamics for *infinitesimal deformations*

$$\nabla \cdot \boldsymbol{\sigma} + \boldsymbol{f} = \rho_o \frac{d^2 \boldsymbol{u}}{dt^2} = \rho_o \frac{d\boldsymbol{v}}{dt}, \tag{9.1}$$

where $\nabla = \nabla_X$ and $\frac{d}{dt} = \frac{\partial}{\partial t}$ (see Appendix B).

9.2 Generic Time Stepping

In order to motivate the time-stepping process, we first start with the dynamics of single point mass under the action of a force $\boldsymbol{\Psi}$. The equation of motion is given by (Newton's Law)

$$m\dot{\boldsymbol{v}} = \boldsymbol{\Psi}, \tag{9.2}$$

where $\boldsymbol{\Psi}$ is the total force applied to the particle. Expanding the velocity in a Taylor series about $t + \theta \Delta t$, where $0 \leq \theta \leq 1$, for $\boldsymbol{v}(t + \Delta t)$, we obtain

$$\boldsymbol{v}(t + \Delta t) = \boldsymbol{v}(t + \theta \Delta t) + \frac{d\boldsymbol{v}}{dt}|_{t+\theta\Delta t}(1-\theta)\Delta t + \frac{1}{2}\frac{d^2\boldsymbol{v}}{dt^2}|_{t+\theta\Delta t}(1-\theta)^2(\Delta t)^2 + \mathcal{O}(\Delta t)^3 \tag{9.3}$$

and for $\boldsymbol{v}(t)$, we obtain

$$\boldsymbol{v}(t) = \boldsymbol{v}(t + \theta \Delta t) - \frac{d\boldsymbol{v}}{dt}|_{t+\theta\Delta t}\theta\Delta t + \frac{1}{2}\frac{d^2\boldsymbol{v}}{dt^2}|_{t+\theta\Delta t}\theta^2(\Delta t)^2 + \mathcal{O}(\Delta t)^3. \tag{9.4}$$

T. I. Zohdi, *A Finite Element Primer for Beginners*, The Basics,
https://doi.org/10.1007/978-3-319-70428-9_9

Subtracting the two expressions yields

$$\frac{d\boldsymbol{v}}{dt}|_{t+\theta\Delta t} = \frac{\boldsymbol{v}(t+\Delta t) - \boldsymbol{v}(t)}{\Delta t} + \hat{\mathcal{O}}(\Delta t), \tag{9.5}$$

where $\hat{\mathcal{O}}(\Delta t) = \mathcal{O}(\Delta t)^2$, when $\theta = \frac{1}{2}$, otherwise $\hat{\mathcal{O}}(\Delta t) = \mathcal{O}(\Delta t)$. Thus, inserting this into Eq. 9.2 yields

$$\boldsymbol{v}(t+\Delta t) = \boldsymbol{v}(t) + \frac{\Delta t}{m}\boldsymbol{\Psi}(t+\theta\Delta t) + \hat{\mathcal{O}}(\Delta t)^2. \tag{9.6}$$

Note that a weighted sum of Eqs. 9.3 and 9.4 yields

$$\boldsymbol{v}(t+\theta\Delta t) = \theta\boldsymbol{v}(t+\Delta t) + (1-\theta)\boldsymbol{v}(t) + \mathcal{O}(\Delta t)^2, \tag{9.7}$$

which will be useful shortly. Now expanding the position of the mass in a Taylor series about $t+\theta\Delta t$ we obtain

$$\boldsymbol{u}(t+\Delta t) = \boldsymbol{u}(t+\theta\Delta t) + \frac{d\boldsymbol{u}}{dt}|_{t+\theta\Delta t}(1-\theta)\Delta t + \frac{1}{2}\frac{d^2\boldsymbol{u}}{dt^2}|_{t+\theta\Delta t}(1-\theta)^2(\Delta t)^2 + \mathcal{O}(\Delta t)^3 \tag{9.8}$$

and

$$\boldsymbol{u}(t) = \boldsymbol{u}(t+\theta\Delta t) - \frac{d\boldsymbol{u}}{dt}|_{t+\theta\Delta t}\theta\Delta t + \frac{1}{2}\frac{d^2\boldsymbol{u}}{dt^2}|_{t+\theta\Delta t}\theta^2(\Delta t)^2 + \mathcal{O}(\Delta t)^3. \tag{9.9}$$

Subtracting the two expressions yields

$$\frac{\boldsymbol{u}(t+\Delta t) - \boldsymbol{u}(t)}{\Delta t} = \boldsymbol{v}(t+\theta\Delta t) + \hat{\mathcal{O}}(\Delta t). \tag{9.10}$$

Inserting Eq. 9.7 yields

$$\boldsymbol{u}(t+\Delta t) = \boldsymbol{u}(t) + (\theta\boldsymbol{v}(t+\Delta t) + (1-\theta)\boldsymbol{v}(t))\,\Delta t + \hat{\mathcal{O}}(\Delta t)^2, \tag{9.11}$$

and using Eq. 9.6 yields

$$\boldsymbol{u}(t+\Delta t) = \boldsymbol{u}(t) + \boldsymbol{v}(t)\Delta t + \frac{\theta(\Delta t)^2}{m}\boldsymbol{\Psi}(t+\theta\Delta t) + \hat{\mathcal{O}}(\Delta t)^2. \tag{9.12}$$

The term $\boldsymbol{\Psi}(t+\theta\Delta t)$ can be handled in a simple way:

$$\boldsymbol{\Psi}(t+\theta\Delta t) \approx \theta\boldsymbol{\Psi}(t+\Delta t) + (1-\theta)\boldsymbol{\Psi}(t). \tag{9.13}$$

We note that

- When $\theta = 1$, then this is the (implicit) Backward Euler scheme, which is very stable (very dissipative) and $\hat{\mathcal{O}}(\Delta t)^2 = \mathcal{O}(\Delta t)^2$ locally in time,
- When $\theta = 0$, then this is the (explicit) Forward Euler scheme, which is conditionally stable and $\hat{\mathcal{O}}(\Delta t)^2 = \mathcal{O}(\Delta t)^2$ locally in time,
- When $\theta = 0.5$, then this is the (implicit) "Midpoint" scheme, which is stable and $\hat{\mathcal{O}}(\Delta t)^2 = \mathcal{O}(\Delta t)^3$ locally in time.

In summary, we have for the velocity[1]

$$\boldsymbol{v}(t + \Delta t) = \boldsymbol{v}(t) + \frac{\Delta t}{m} \left(\theta \boldsymbol{\Psi}(t + \Delta t) + (1 - \theta)\boldsymbol{\Psi}(t)\right) \tag{9.14}$$

and for the position

$$\begin{aligned} \boldsymbol{u}(t + \Delta t) &= \boldsymbol{u}(t) + \boldsymbol{v}(t + \theta \Delta t)\Delta t \\ &= \boldsymbol{u}(t) + \left(\theta \boldsymbol{v}(t + \Delta t) + (1 - \theta) bf\, v(t)\right) \Delta t, \end{aligned} \tag{9.15}$$

or in terms of $\boldsymbol{\Psi}$

$$\boldsymbol{u}(t + \Delta t) = \boldsymbol{u}(t) + \boldsymbol{v}(t)\Delta t + \frac{\theta(\Delta t)^2}{m} \left(\theta \boldsymbol{\Psi}(t + \Delta t) + (1 - \theta)\boldsymbol{\Psi}(t)\right). \tag{9.16}$$

9.3 Application to the Continuum Formulation

Now consider the continuum analogue to "$m\dot{\boldsymbol{v}}$"

$$\rho_o \frac{\partial^2 \boldsymbol{u}}{\partial t^2} = \rho_o \frac{\partial \boldsymbol{v}}{\partial t} = \nabla \cdot \boldsymbol{\sigma} + \boldsymbol{f} \stackrel{\text{def}}{=} \boldsymbol{\Psi} \tag{9.17}$$

and thus

$$\rho_o \boldsymbol{v}(t + \Delta t) = \rho_o \boldsymbol{v}(t) + \Delta t \left(\theta \boldsymbol{\Psi}(t + \Delta t) + (1 - \theta)\boldsymbol{\Psi}(t)\right). \tag{9.18}$$

Multiplying Eq. 9.18 by a test function and integrating yields

$$\begin{aligned} \int_\Omega \boldsymbol{\nu} \cdot \rho_o \boldsymbol{v}(t + \Delta t)\, d\Omega &= \int_\Omega \boldsymbol{\nu} \cdot \rho_o \boldsymbol{v}(t)\, d\Omega \\ &\quad + \Delta t \int_\Omega \boldsymbol{\nu} \cdot \left(\theta \boldsymbol{\Psi}(t + \Delta t) + (1 - \theta)\boldsymbol{\Psi}(t)\right) d\Omega, \end{aligned} \tag{9.19}$$

[1] In order to streamline the notation, we drop the cumbersome $\mathcal{O}(\Delta t)$-type terms.

and using Gauss's divergence theorem and enforcing $\boldsymbol{v} = \boldsymbol{0}$ on Γ_u yields (using a streamlined time-step superscript counter notation of L, where $t = L\Delta t$ and $t + \Delta t = (L+1)\Delta t$)

$$\int_{\Omega} \boldsymbol{v} \cdot \rho_o \boldsymbol{v}^{L+1}\, d\Omega = \int_{\Omega} \boldsymbol{v} \cdot \rho_o \boldsymbol{v}^{L}\, d\Omega \tag{9.20}$$
$$+ \Delta t\theta \left(- \int_{\Omega} \nabla \boldsymbol{v} : \boldsymbol{\sigma}\, d\Omega + \int_{\Gamma_t} \boldsymbol{v} \cdot (\boldsymbol{\sigma} \cdot \boldsymbol{n})\, dA + \int_{\Omega} \boldsymbol{v} \cdot \boldsymbol{f}\, d\Omega \right)^{L+1}$$
$$+ \Delta t(1-\theta) \left(- \int_{\Omega} \nabla \boldsymbol{v} : \boldsymbol{\sigma}\, d\Omega + \int_{\Gamma_t} \boldsymbol{v} \cdot \boldsymbol{t}^*\, dA + \int_{\Omega} \boldsymbol{v} \cdot \boldsymbol{f}\, d\Omega \right)^{L}.$$

As in the previous chapter on linearized three-dimensional elasticity, we assume

$$\{\boldsymbol{u}^h\} = [\boldsymbol{\Phi}]\{\boldsymbol{a}\} \quad \textit{and} \quad \{\boldsymbol{v}^h\} = [\boldsymbol{\Phi}]\{\boldsymbol{b}\} \quad \textit{and} \quad \{\boldsymbol{v}^h\} = [\boldsymbol{\Phi}]\{\dot{\boldsymbol{a}}\}, \tag{9.21}$$

which yields, in terms of matrices and vectors

$$\{\boldsymbol{b}\}^T[M]\{\dot{\boldsymbol{a}}\}^{L+1} = \{\boldsymbol{b}\}^T[M]\{\dot{\boldsymbol{a}}\}^{L} - \Delta t\theta\{\boldsymbol{b}\}^T \left(-[K]\{\boldsymbol{a}\}^{L+1} + \{\boldsymbol{R}_f\}^{L+1} + \{\boldsymbol{R}_t\}^{L+1} \right)$$
$$- \{\boldsymbol{b}\}^T \Delta t(1-\theta) \left(-[K]\{\boldsymbol{a}\}^{L} + \{\boldsymbol{R}_f\}^{L} + \{\boldsymbol{R}_t\}^{L} \right). \tag{9.22}$$

where $[M] = \int_{\Omega} \rho_o [\Phi]^T [\Phi]\, d\Omega$, and $[K]$,$\{\boldsymbol{R}_f\}$, and $\{\boldsymbol{R}_t\}$ are as defined in the previous chapters on elastostatics. Note that $\{\boldsymbol{R}_f\}^L$ and $\{\boldsymbol{R}_t\}^L$ are known values from the previous time-step. Since $\{\boldsymbol{b}\}^T$ is arbitrary

$$[M]\{\dot{\boldsymbol{a}}\}^{L+1} = [M]\{\dot{\boldsymbol{a}}\}^{L} + (\Delta t\theta) \left(-[K]\{\boldsymbol{a}\}^{L+1} + \{\boldsymbol{R}_f\}^{L+1} + \{\boldsymbol{R}_t\}^{L+1} \right)$$
$$+ \Delta t(1-\theta) \left(-[K]\{\boldsymbol{a}\}^{L} + \{\boldsymbol{R}_f\}^{L} + \{\boldsymbol{R}_t\}^{L} \right). \tag{9.23}$$

One should augment this with the approximation for the discrete displacement:

$$\{\boldsymbol{a}\}^{L+1} = \{\boldsymbol{a}\}^{L} + \Delta t \left(\theta\{\dot{\boldsymbol{a}}\}^{L+1} + (1-\theta)\{\dot{\boldsymbol{a}}\}^{L} \right). \tag{9.24}$$

For a purely implicit (Backward Euler) method $\theta = 1$

$$\left([M]\{\dot{\boldsymbol{a}}\}^{L+1} + \Delta t[K]\{\boldsymbol{a}\}^{L+1} \right) = [M]\{\dot{\boldsymbol{a}}\}^{L} + \Delta t \left(\{\boldsymbol{R}_t\}^{L+1} + \{\boldsymbol{R}_f\}^{L+1} \right), \tag{9.25}$$

augmented with

$$\{\boldsymbol{a}\}^{L+1} = \{\boldsymbol{a}\}^{L} + \Delta t\{\dot{\boldsymbol{a}}\}^{L+1}, \tag{9.26}$$

which requires one to solve a system of algebraic equations, while for an explicit (Forward Euler) method $\theta = 0$ with usually $[M]$ is approximated by an easy-to-invert matrix, such as a diagonal matrix, $[M] \approx M[\mathbf{1}]$, to make the matrix inversion easy, yielding:

$$\{\dot{\boldsymbol{a}}\}^{L+1} = \{\dot{\boldsymbol{a}}\}^{L} + \Delta t[M]^{-1} \left(-[K]\{\boldsymbol{a}\}^{L} + \{\boldsymbol{R}_f\}^{L} + \{\boldsymbol{R}_t\}^{L} \right), \tag{9.27}$$

augmented with

$$\{a\}^{L+1} = \{a\}^L + \Delta t\{\dot{a}\}^L. \tag{9.28}$$

There is an enormous number of time-stepping schemes. For general time-stepping, we refer the reader to the seminal texts of Hairer et al. [1,2]. In the finite element context, we refer the reader to Bathe [3], Becker et al. [4], Hughes [5], and Zienkiewicz and Taylor [6].

References

1. Hairer, E., Norsett, S. P., & Wanner, G. (2000). *Solving ordinary differential equations I. Nonstiff equations* (2nd ed.). Heidelberg: Springer.
2. Hairer, E., Lubich, C., & Wanner, G. (2006). *Solving ordinary differential equations II. Stiff and differential-algebraic problems* (2nd ed.). Heidelberg: Springer.
3. Bathe, K. J. (1996). *Finite element procedures*. Englewood Cliffs: Prentice Hall.
4. Becker, E. B., Carey, G. F., & Oden, J. T. (1980). *Finite elements: An introduction*. Englewood Cliffs: Prentice Hall.
5. Hughes, T. J. R. (1989). *The finite element method*. Englewood Cliffs: Prentice Hall.
6. Zienkiewicz, O. C., & Taylor, R. L. (1991). *The finite element method* (Vol. I and II). New York: McGraw-Hill.

Summary and Advanced Topics 10

The finite element method is a huge field of study. This set of notes was designed to give students only a brief introduction to the fundamentals of the method. The implementation, theory, and application of FEM is a subject of immense literature. For general references on the subject, see the well-known books of Ainsworth and Oden [1], Becker et al. [2], Carey and Oden [3], Oden and Carey [4], Hughes [5], Szabo and Babuska [6], Bathe [7], and Zienkiewicz and Taylor [8]. For a review of the state of the art in finite element methods, see the relatively recent book of Wriggers [9]. Much of the modern research activity in computational mechanics reflects the growing industrial demands for rapid simulation of large-scale, nonlinear, time-dependent problems. Accordingly, the next concepts the reader should focus on are:

1. Error estimation and adaptive mesh refinement,
2. Time-dependent problems,
3. Geometrically and materially nonlinear problems and
4. High-performance computing: domain decomposition and parallel processing.

The last item is particularly important. Thus, we close with a few comments on domain decomposition and parallel processing.

In many cases, in particular in three dimensions, for a desired accuracy, the meshes need to be so fine that the number of unknowns outstrips the available computing power on a single serial processing machine. One approach to deal with this problem is domain decomposition. Decomposition of a domain into parts (subdomains) that can be solved independently by estimating the boundary conditions, solving the decoupled subdomains, correcting the boundary conditions by updating them using information from the computed solutions, and repeating the procedure has become popular over the last 20 years as a means of harnessing computational power afforded by parallel processing machines.

T. I. Zohdi, *A Finite Element Primer for Beginners*, The Basics,
https://doi.org/10.1007/978-3-319-70428-9_10

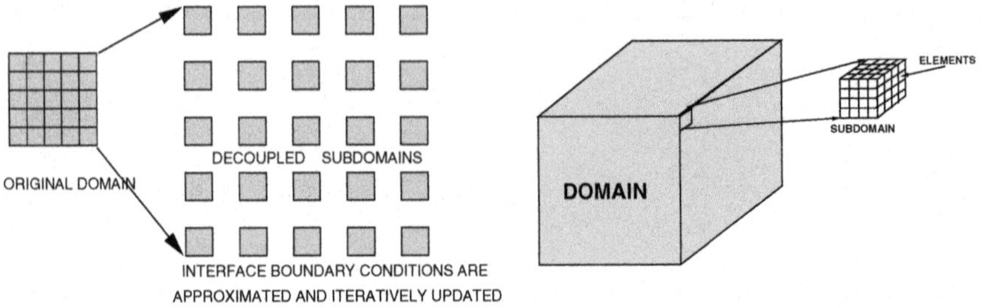

Fig. 10.1 Left: A two-dimensional view of the decomposition of a domain and Right: a three-dimensional view

Consider the three-dimensional block (an elasticity problem) where we use linear brick elements with the following parameters (Fig. 10.1):

- The number of subdomains: $M \times M \times M$.
- The number elements in each subdomain: $N \times N \times N$.

For the original (nonpartitioned domain):

- The number elements: $(N \times M) \times (N \times M) \times (N \times M)$.
- The number nodes: $(N \times M + 1) \times (N \times M + 1) \times (N \times M + 1)$.
- The number degrees of freedom (for elasticity): $3 \times (N \times M + 1) \times (N \times M + 1) \times (N \times M + 1)$.
- The number elements in the entire decoupled domain: $(N \times M) \times (N \times M) \times (N \times M)$.
- The data storage for the entire domain: $(N \times M)^3 \times 300$ (symmetric storage for elasticity).

For the partitioned domain:

- The number nodes in each subdomain: $(N + 1) \times (N + 1) \times (N + 1)$.
- The number degrees of freedom (for elasticity) in each subdomain: $3 \times (N + 1) \times (N + 1) \times (N + 1)$.
- The data storage per subdomain: $N^3 \times 300$ (symmetric storage for elasticity).

Let us now consider:

- The number processors involved: P.
- The number iterations needed to update the interface solution: I.

The operation counts for solving the whole domain is

$$C_d \propto ((3(NM + 1))^3)^\gamma, \tag{10.1}$$

while for each subdomain

$$C_{sd} \propto ((3(N+1))^3)^\gamma, \tag{10.2}$$

where $1 \leq \gamma \leq 3$ is an exponent that reflects the extremes of solving efficiency. The ratio of the amount of work done by solving the total domain to that of solving the subdomain problems (taking into account the number of iterations (I) needed to update the interface boundary conditions) is approximately

$$\frac{C_d}{IC_{sd}} \propto \frac{((3(NM+1))^3)^\gamma}{I((3(N+1))^3)^\gamma} \approx \frac{M^{3\gamma}}{I}, \tag{10.3}$$

where we have ignored the costs of computing the updated interface conditions (considered small). If we assume that the amount of time to solve is also proportional to the operation counts, and assume that each domain is processed in the same amount of time, using P processors yields:

$$\frac{C_d}{IC_{sd}/P} = \frac{PM^{3\gamma}}{I}. \tag{10.4}$$

In order to understand the scaling numerically, consider

- One-thousand processors: $P = 10^3$,
- One-thousand subdomains: $M \times M \times M = 10 \times 10 \times 10$.
- The number of updates: $I = 10^2$.

The resulting ratio of computational costs is:

- For $\gamma = 3 : \frac{C_d}{IC_{sd}/P} = 10^{10}$,
- For $\gamma = 2 : \frac{C_d}{IC_{sd}/P} = 10^{7}$.
- For $\gamma = 1 : \frac{C_d}{IC_{sd}/P} = 10^{4}$.

This idealized simple example illustrates the possible benefits in reduction of solution time, independent of the gains in data storage. For a historical overview, as well as a thorough analysis of the wide range of approaches, see Le Tallec [10]. In many cases, interprocessor communication and synchronization can be a bottleneck to obtain a high-performance parallel algorithm. The parallel speedup (relative to a sequential implementation), S, can be approximated by Amdahl's law (Amdahl [11]), $S = \frac{1}{1-f}$, where f is the fraction of the algorithm that is parallelizable. For example, if 40% of the code is inherently sequential, then $f = 0.6$ and $S = 2.5$. This provides an upper bound on the utility of adding more processors. A related expression is "Gustafson's law" Gustafson [12], $S(f) = f - k(f - 1)$, where k represents the parts of the algorithm that are not parallelizable. Amdahl's law assumes that the problem is of fixed size and that the sequential part is independent of the number of processors; however, Gustafson's law does not make either of these assumptions. We refer the reader to the works of Papadrakakis et al. [13,14] for parallel strategies that are directly applicable to the class of problems of interest.

Remarks: Some comments of the convergence of such iterative schemes are provided in Appendix C.

References

1. Ainsworth, M., & Oden, J. T. (2000). *A posterori error estimation in finite element analysis*. New York: Wiley.
2. Becker, E. B., Carey, G. F., & Oden, J. T. (1980). *Finite elements: An introduction*. Englewood Cliffs: Prentice Hall.
3. Carey, G. F., & Oden, J. T. (1983). *Finite elements: A second course*. Englewood Cliffs: Prentice Hall.
4. Oden, J. T., & Carey, G. F. (1984). *Finite elements: Mathematical aspects*. Englewood Cliffs: Prentice Hall.
5. Hughes, T. J. R. (1989). *The finite element method*. Englewood Cliffs: Prentice Hall.
6. Szabo, B., & Babúska, I. (1991). *Finite element analysis*. New York: Wiley Interscience.
7. Bathe, K. J. (1996). *Finite element procedures*. Englewood Cliffs: Prentice Hall.
8. Zienkiewicz, O. C., & Taylor, R. L. (1991). *The finite element method* (Vol. I and II). New York: McGraw-Hill.
9. Wriggers, P. (2008). *Nonlinear finite element analysis*. Berlin: Springer.
10. Le Tallec, P. (1994). Domain decomposition methods in computational mechanics. *Computational Mechanics Advances, 1*, 121–220.
11. Amdahl, G. (1967). The validity of a single processor approach to achieving large-scale computing capabilities. In *Proceedings of AFIPS Spring Joint Computer Conference* (pp. 483–485). Atlantic City, N. J.: AFIPS Press.
12. Gustafson, J. L. (1988). Reevaluating Amdahl's law. *Communications of the ACM, 31*(5), 532–533.
13. Papadrakakis, M. (1993). *Solving large-scale problems in mechanics*. New York: Wiley.
14. Papadrakakis, M. (1997). *Parallel solution methods in computational mechanics*. Chichester: Wiley.

Appendix A Elementary Mathematical Concepts

A

Throughout this document, boldface symbols imply vectors or tensors (matrices in our analyses).

A.1 Vector Products

For the inner product of two vectors (first-order tensors) $\boldsymbol{u}$ and $\boldsymbol{v}$ we have in three dimensions

$$\boldsymbol{u} \cdot \boldsymbol{v} = \underbrace{u_i v_i}_{\text{in Cartesian bases}} = u_1 v_1 + u_2 v_2 + u_3 v_3 = |\boldsymbol{u}||\boldsymbol{v}| cos\theta, \tag{A.1}$$

where $|\boldsymbol{u}| = \sqrt{u_1^2 + u_2^2 + u_3^2}$ and where Einstein index summation notation is used. Two vectors are said to be orthogonal if $\boldsymbol{u} \cdot \boldsymbol{v} = 0$. The cross (vector) product of two vectors is

$$\boldsymbol{u} \times \boldsymbol{v} = \left(\begin{vmatrix} e_1 & e_2 & e_3 \\ u_1 & u_2 & u_3 \\ v_1 & v_2 & v_3 \end{vmatrix} \right) = |\boldsymbol{u}||\boldsymbol{v}| sin\theta \, \boldsymbol{n}, \tag{A.2}$$

where $\boldsymbol{n}$ is the unit normal to the plane formed by the vectors $\boldsymbol{u}$ and $\boldsymbol{v}$. The triple product of three vectors is

$$\boldsymbol{w} \cdot (\boldsymbol{u} \times \boldsymbol{v}) = \left(\begin{vmatrix} w_1 & w_2 & w_3 \\ u_1 & u_2 & u_3 \\ v_1 & v_2 & v_3 \end{vmatrix} \right) = (\boldsymbol{w} \times \boldsymbol{u}) \cdot \boldsymbol{v} \tag{A.3}$$

This represents the volume of a parallelepiped formed by the three vectors.

T. I. Zohdi, *A Finite Element Primer for Beginners*, The Basics,
https://doi.org/10.1007/978-3-319-70428-9

A.2 Vector Calculus

We have the following elementary operations:

- The divergence of a vector (a contraction to a scalar) is defined by

$$\nabla \cdot \boldsymbol{u} = u_{i,i} \tag{A.4}$$

  whereas for a second-order tensor (a contraction to a vector):

$$\nabla \cdot \boldsymbol{A} \text{ has components of } A_{ij,j}. \tag{A.5}$$

- The gradient of a vector (a dilation to a second-order tensor) is:

$$\nabla \boldsymbol{u} \text{ has components of } u_{i,j}, \tag{A.6}$$

  whereas for a second-order tensor (a dilation to a third-order tensor):

$$\nabla \boldsymbol{A} \text{ has components of } A_{ij,k}. \tag{A.7}$$

- The gradient of a scalar (a dilation to a vector) is:

$$\nabla \phi \text{ has components of } \phi_{,i}. \tag{A.8}$$

  The scalar product of two second-order tensors, for example, the gradients of first-order vectors, is defined as

$$\nabla \boldsymbol{v} : \nabla \boldsymbol{u} = \underbrace{\frac{\partial v_i}{\partial x_j}\frac{\partial u_i}{\partial x_j}}_{\text{in Cartesian bases}} \stackrel{\text{def}}{=} v_{i,j}u_{i,j} \qquad i, j = 1, 2, 3, \tag{A.9}$$

  where $\partial u_i/\partial x_j$, $\partial v_i/\partial x_j$ are partial derivatives of u_i and v_i, and where u_i, v_i are the Cartesian components of $\boldsymbol{u}$ and $\boldsymbol{v}$ and

$$\nabla \boldsymbol{u} \cdot \boldsymbol{n} \text{ has components of } \underbrace{u_{i,j}n_j}_{\text{in Cartesian bases}}, \qquad i, j = 1, 2, 3. \tag{A.10}$$

- The divergence theorem for vectors is

$$\int_{\Omega} \nabla \cdot \boldsymbol{u}\, d\Omega = \int_{\partial\Omega} \boldsymbol{u} \cdot \boldsymbol{n}\, dA \qquad \int_{\Omega} u_{i,i}\, d\Omega = \int_{\partial\Omega} u_i n_i\, dA \tag{A.11}$$

  and analogously for tensors

$$\int_{\Omega} \nabla \cdot \boldsymbol{B}\, d\Omega = \int_{\partial\Omega} \boldsymbol{B} \cdot \boldsymbol{n}\, dA \qquad \int_{\Omega} B_{ij,j}\, d\Omega = \int_{\partial\Omega} B_{ij} n_j\, dA, \tag{A.12}$$

  where $\boldsymbol{n}$ is the outward normal to the bounding surface.

These standard operations arise throughout the analysis.

A.3 Interpretation of the Gradient of Functionals

The elementary concepts to follow are important for understanding iterative solvers. Consider a surface in space defined by

$$\Pi(x_1, x_2, ...x_N) = C. \tag{A.13}$$

Consider a unit vector $\boldsymbol{b}$, and the inner product, forming the directional derivative (the rate of change of Π in the direction of $\boldsymbol{b}$):

$$\nabla\Pi \cdot \boldsymbol{b} = ||\boldsymbol{b}||||\nabla\Pi||cos\gamma. \tag{A.14}$$

When $\gamma = 0$, the directional derivative is maximized, in other words when $\boldsymbol{b}$ and $\nabla\Pi$ are colinear. Since we can represent curves on the surface defined by $\Pi = C$ by a position vector (t is a parameter)

$$\boldsymbol{r} = x_1(t)\boldsymbol{e}_1 + x_2(t)\boldsymbol{e}_2... + x_N(t)\boldsymbol{e}_N, \tag{A.15}$$

the tangent is

$$\frac{d\boldsymbol{r}}{dt} = \frac{dx_1}{dt}\boldsymbol{e}_1 + \frac{dx_2}{dt}\boldsymbol{e}_2... + \frac{dx_N}{dt}\boldsymbol{e}_N. \tag{A.16}$$

If we take

$$\frac{d\Pi}{dt} = 0 = \nabla\Pi \cdot \frac{d\boldsymbol{r}}{dt} = \frac{\partial\Pi}{\partial x_1}\frac{dx_1}{dt} + \frac{\partial\Pi}{\partial x_2}\frac{dx_2}{dt}... + \frac{\partial\Pi}{\partial x_N}\frac{dx_N}{dt}, \tag{A.17}$$

we immediately see that $\nabla\Pi$ is normal to the surface and represents the direction of maximum change in the normal direction.

A.4 Matrix Manipulations

Throughout the next few definitions, we consider the matrix $[\boldsymbol{A}]$. The matrix $[\boldsymbol{A}]$ is said to be symmetric if $[\boldsymbol{A}] = [\boldsymbol{A}]^T$ and skew-symmetric if $[\boldsymbol{A}] = -[\boldsymbol{A}]^T$. A first-order contraction (inner product) of two matrices is defined by

$$\boldsymbol{A} \cdot \boldsymbol{B} = [\boldsymbol{A}][\boldsymbol{B}] \text{ has components of } A_{ij}B_{jk} = C_{ik} \tag{A.18}$$

where it is clear that the range of the inner index j must be the same for $[\boldsymbol{A}]$ and $[\boldsymbol{B}]$. The second-order inner product of two matrices is

$$\boldsymbol{A} : \boldsymbol{B} = A_{ij}B_{ij} = tr([\boldsymbol{A}]^T[\boldsymbol{B}]) \tag{A.19}$$

The rule of transposes for the product of two matrices is

$$([\boldsymbol{A}][\boldsymbol{B}])^T = [\boldsymbol{B}]^T[\boldsymbol{A}]^T. \tag{A.20}$$

The rule of inverses for two invertible $n \times n$ matrices is

$$([\boldsymbol{A}][\boldsymbol{B}])^{-1} = [\boldsymbol{B}]^{-1}[\boldsymbol{A}]^{-1} \qquad [\boldsymbol{A}]^{-1}[\boldsymbol{A}] = [\boldsymbol{A}][\boldsymbol{A}]^{-1} = [\boldsymbol{1}] \tag{A.21}$$

where $[\boldsymbol{1}]$ is the identity matrix. Clearly, $[\boldsymbol{A}]^{-1}$ exists only when $det[\boldsymbol{A}] \neq 0$.

A.4.1 Determinant

Some properties of the determinant (where $[\boldsymbol{A}]$ is a 3×3 matrix):

$$[\boldsymbol{A}] \stackrel{\text{def}}{=} \begin{bmatrix} A_{11} & A_{12} & A_{13} \\ A_{21} & A_{22} & A_{23} \\ A_{31} & A_{32} & A_{33} \end{bmatrix} \tag{A.22}$$

are

$$\boxed{\begin{array}{l} det[\boldsymbol{A}] = A_{11}(A_{22}A_{33} - A_{32}A_{23}) - A_{12}(A_{21}A_{33} - A_{31}A_{23}) + A_{13}(A_{21}A_{32} - A_{31}A_{22}), \\ det[\boldsymbol{1}] = 1, \qquad det\,\alpha[\boldsymbol{A}] = \alpha^3 det\,[\boldsymbol{A}], \qquad \alpha = scalar, \\ det[\boldsymbol{A}][\boldsymbol{B}] = det[\boldsymbol{A}]det[\boldsymbol{B}], \qquad det[\boldsymbol{A}]^T = det[\boldsymbol{A}], \qquad det[\boldsymbol{A}]^{-1} = \frac{1}{det[\boldsymbol{A}]}. \end{array}}$$

An important use of the determinant is in forming the inverse by

$$[\boldsymbol{A}]^{-1} = \frac{adj[\boldsymbol{A}]}{det[\boldsymbol{A}]}, \qquad adj[\boldsymbol{A}] \stackrel{\text{def}}{=} \begin{bmatrix} C_{11} & C_{12} & C_{13} \\ C_{21} & C_{22} & C_{23} \\ C_{31} & C_{32} & C_{33} \end{bmatrix}^T, \tag{A.23}$$

where the so-called cofactors are

$$\boxed{\begin{array}{ll} C_{11} = A_{22}A_{33} - A_{32}A_{23} & C_{12} = -(A_{21}A_{33} - A_{31}A_{23}) \\ C_{13} = A_{21}A_{32} - A_{31}A_{22} & C_{21} = -(A_{12}A_{33} - A_{32}A_{13}) \\ C_{22} = A_{11}A_{33} - A_{31}A_{13} & C_{23} = -(A_{11}A_{32} - A_{31}A_{12}) \\ C_{31} = A_{12}A_{23} - A_{22}A_{13} & C_{32} = -(A_{11}A_{23} - A_{21}A_{13}) \\ C_{33} = A_{11}A_{22} - A_{21}A_{12} & \end{array}} \tag{A.24}$$

A.4.2 Eigenvalues

The mathematical definition of an eigenvalue, a scalar denoted Λ, and eigenvector, a vector denoted $\mathcal{E}$, of a matrix $[\boldsymbol{A}]$ is

$$[\boldsymbol{A}]\{\mathcal{E}\} = \Lambda\{\mathcal{E}\} \tag{A.25}$$

Some main properties to remember about eigenvalues and eigenvectors are:

1. If $[\boldsymbol{A}]$ $(n \times n)$ has n linearly independent eigenvectors then it is diagonalizable by a matrix formed by columns of the eigenvectors. In the case of a 3×3 matrix,

$$\begin{bmatrix} \Lambda_1 & 0 & 0 \\ 0 & \Lambda_2 & 0 \\ 0 & 0 & \Lambda_3 \end{bmatrix} = \begin{bmatrix} \mathcal{E}_1^{(1)} & \mathcal{E}_1^{(2)} & \mathcal{E}_1^{(3)} \\ \mathcal{E}_2^{(1)} & \mathcal{E}_2^{(2)} & \mathcal{E}_2^{(3)} \\ \mathcal{E}_3^{(1)} & \mathcal{E}_3^{(2)} & \mathcal{E}_3^{(3)} \end{bmatrix}^{-1} \begin{bmatrix} A_{11} & A_{12} & A_{13} \\ A_{21} & A_{22} & A_{23} \\ A_{31} & A_{32} & A_{33} \end{bmatrix} \begin{bmatrix} \mathcal{E}_1^{(1)} & \mathcal{E}_1^{(2)} & \mathcal{E}_1^{(3)} \\ \mathcal{E}_2^{(1)} & \mathcal{E}_2^{(2)} & \mathcal{E}_2^{(3)} \\ \mathcal{E}_3^{(1)} & \mathcal{E}_3^{(2)} & \mathcal{E}_3^{(3)} \end{bmatrix} \tag{A.26}$$

2. If $[\boldsymbol{A}]$ $(n \times n)$ has n distinct eigenvalues then the eigenvectors are linearly independent
3. If $[\boldsymbol{A}]$ $(n \times n)$ is symmetric then its eigenvalues are real. If the eigenvalues are distinct, then the eigenvectors are orthogonal.

A quadratic form is defined as $\{\boldsymbol{x}\}^T[\boldsymbol{A}]\{\boldsymbol{x}\}$, and is positive when $[\boldsymbol{A}]$ has positive eigenvalues. Explicitly, for a 3×3 matrix, we have

$$\{\boldsymbol{x}\}^T[\boldsymbol{A}]\{\boldsymbol{x}\} \stackrel{\text{def}}{=} \begin{bmatrix} x_1 & x_2 & x_3 \end{bmatrix} \begin{bmatrix} A_{11} & A_{12} & A_{13} \\ A_{21} & A_{22} & A_{23} \\ A_{31} & A_{32} & A_{33} \end{bmatrix} \begin{bmatrix} x_1 \\ x_2 \\ x_3 \end{bmatrix}. \tag{A.27}$$

A matrix $[\boldsymbol{A}]$ is said to be positive definite if the quadratic form is positive for all nonzero vectors $\boldsymbol{x}$. Clearly, from Eq. A.26 a positive definite matrix must have positive eigenvalues.

Remark: If we set the determinant $det[\boldsymbol{A} - \Lambda\boldsymbol{1}] = 0$, it can be shown that the so-called characteristic polynomial is, for example for a 3×3 matrix:

$$det(\boldsymbol{A} - \Lambda\boldsymbol{1}) = -\Lambda^3 + I_A\Lambda^2 - II_A\Lambda + III_A = 0, \tag{A.28}$$

where

$$\boxed{\begin{aligned} I_A &= tr(\boldsymbol{A}) = \Lambda_1 + \Lambda_2 + \Lambda_3 \\ II_A &= \tfrac{1}{2}((tr(\boldsymbol{A}))^2 - tr(\boldsymbol{A}^2)) = \Lambda_1\Lambda_2 + \Lambda_2\Lambda_3 + \Lambda_1\Lambda_3 \\ III_A &= det(\boldsymbol{A}) = \tfrac{1}{6}((tr\boldsymbol{A})^3 + 2tr\boldsymbol{A}^3 - 3(tr\boldsymbol{A}^2)(tr\boldsymbol{A})) = \Lambda_1\Lambda_2\Lambda_3. \end{aligned}} \tag{A.29}$$

Since I_A, II_A, and III_A can be written in terms of $tr\boldsymbol{A}$, which is invariant under frame rotation, they too are invariant under frame rotation.

A.4.3 Coordinate Transformations

To perform a coordinate transform for a 3×3 matrix $[\boldsymbol{A}]$ from one Cartesian coordinate system to another (denoted with a $\hat{(\cdot)}$), we apply a transformation matrix $[\boldsymbol{Q}]$ (Fig. A.1):

$$[\hat{\boldsymbol{A}}] = [\boldsymbol{Q}][\boldsymbol{A}][\boldsymbol{Q}]^{-1} \tag{A.30}$$

In three dimensions, the standard axes rotators are, about the x_1 axis

$$Rot(x_1) \stackrel{\text{def}}{=} \begin{bmatrix} 1 & 0 & 0 \\ 0 & cos\theta_1 & sin\theta_1 \\ 0 & -sin\theta_1 & cos\theta_1 \end{bmatrix}, \tag{A.31}$$

about the x_2 axis

$$Rot(x_2) \stackrel{\text{def}}{=} \begin{bmatrix} cos\theta_2 & 0 & -sin\theta_2 \\ 0 & 1 & 0 \\ sin\theta_2 & 0 & cos\theta_2 \end{bmatrix} \tag{A.32}$$

and about the x_3 axis

$$Rot(x_3) \stackrel{\text{def}}{=} \begin{bmatrix} cos\theta_3 & sin\theta_3 & 0 \\ -sin\theta_3 & cos\theta_3 & 0 \\ 0 & 0 & 1 \end{bmatrix}. \tag{A.33}$$

The standard axes reflectors are, with respect to the $x_2 - x_3$ plane

$$Ref(x_1) \stackrel{\text{def}}{=} \begin{bmatrix} -1 & 0 & 0 \\ 0 & 1 & 0 \\ 0 & 0 & 1 \end{bmatrix}, \tag{A.34}$$

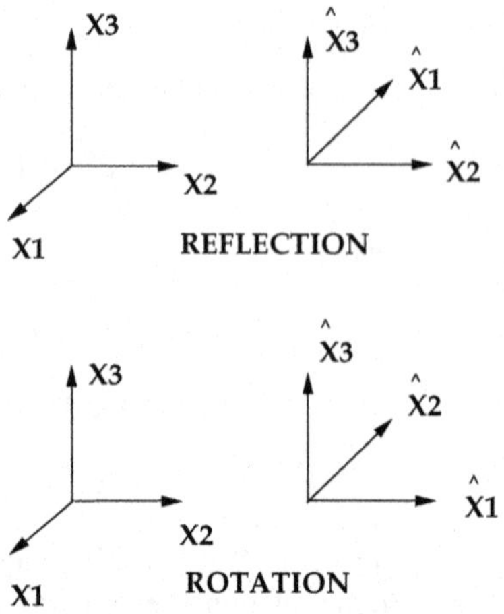

Fig. A.1 Top: reflection with respect to the $x_2 - x_3$ plane. Bottom: rotation with respect to the x_3 axis

with respect to the $x_1 - x_3$ plane

$$Ref(x_2) \stackrel{\text{def}}{=} \begin{bmatrix} 1 & 0 & 0 \\ 0 & -1 & 0 \\ 0 & 0 & 1 \end{bmatrix}, \tag{A.35}$$

with respect to the $x_1 - x_2$ plane

$$Ref(x_3) \stackrel{\text{def}}{=} \begin{bmatrix} 1 & 0 & 0 \\ 0 & 1 & 0 \\ 0 & 0 & -1 \end{bmatrix}. \tag{A.36}$$

Appendix B
Basic Continuum Mechanics

B

In this chapter, we provide the reader with basic background information for field equations of interest.

B.1 Deformations

The term deformation refers to a change in the shape of a continuum between a reference configuration and a current configuration. In the reference configuration, a representative particle of a continuum occupies a point P in space and has the position vector (Fig. B.1)

$$\boldsymbol{X} = X_1 \boldsymbol{e}_1 + X_2 \boldsymbol{e}_2 + X_3 \boldsymbol{e}_3, \tag{B.1}$$

where $\boldsymbol{e}_1, \boldsymbol{e}_2, \boldsymbol{e}_3$ is a Cartesian reference triad, and X_1, X_2, X_3 (with center O) can be thought of as labels for a material point. Sometimes the coordinates or labels (X_1, X_2, X_3) are called the referential or material coordinates. In the current configuration, the particle originally located at point P (at time $t = 0$) is located at point P' and can be also expressed in terms of another position vector $\boldsymbol{x}$, with coordinates (x_1, x_2, x_3). These are called the current coordinates. In this framework, the displacement is $\boldsymbol{u} = \boldsymbol{x} - \boldsymbol{X}$ for a point originally at $\boldsymbol{X}$ and with final coordinates $\boldsymbol{x}$.

When a continuum undergoes deformation (or flow), its points move along various paths in space. This motion may be expressed as a function of $\boldsymbol{X}$ and t as[1]

$$\boldsymbol{x}(\boldsymbol{X}, t) = \boldsymbol{u}(\boldsymbol{X}, t) + \boldsymbol{X}(t), \tag{B.2}$$

[1]Frequently, analysts consider the referential configuration to be fixed in time thus, $\boldsymbol{X} \neq \boldsymbol{X}(t)$. We shall adopt this in the present work.

T. I. Zohdi, *A Finite Element Primer for Beginners*, The Basics,
https://doi.org/10.1007/978-3-319-70428-9

which gives the present location of a point at time t, written in terms of the referential coordinates X_1, X_2, X_3. The previous position vector may be interpreted as a mapping of the initial configuration onto the current configuration. In classical approaches, it is assumed that such a mapping is one-to-one and continuous, with continuous partial derivatives to whatever order required. The description of motion or deformation expressed previously is known as the Lagrangian formulation. Alternatively, if the independent variables are the coordinates $\boldsymbol{x}$ and time t, then $\boldsymbol{x}(x_1, x_2, x_3, t) = \boldsymbol{u}(x_1, x_2, x_3, t) + \boldsymbol{X}(x_1, x_2, x_3, t)$, and the formulation is denoted as Eulerian (Fig. B.1).

Partial differentiation of the displacement vector $\boldsymbol{u} = \boldsymbol{x} - \boldsymbol{X}$, with respect to $\boldsymbol{X}$, produces the following displacement gradient:

$$\nabla_X \boldsymbol{u} = \boldsymbol{F} - \mathbf{1}, \tag{B.3}$$

where

$$\boldsymbol{F} \stackrel{\text{def}}{=} \nabla_X \boldsymbol{x} \stackrel{\text{def}}{=} \frac{\partial \boldsymbol{x}}{\partial \boldsymbol{X}} = \begin{bmatrix} \frac{\partial x_1}{\partial X_1} & \frac{\partial x_1}{\partial X_2} & \frac{\partial x_1}{\partial X_3} \\ \frac{\partial x_2}{\partial X_1} & \frac{\partial x_2}{\partial X_2} & \frac{\partial x_2}{\partial X_3} \\ \frac{\partial x_3}{\partial X_1} & \frac{\partial x_3}{\partial X_2} & \frac{\partial x_3}{\partial X_3} \end{bmatrix}. \tag{B.4}$$

$\boldsymbol{F}$ is known as the material deformation gradient.

Now, consider the length of a differential element in the reference configuration $d\boldsymbol{X}$ and $d\boldsymbol{x}$ in the current configuration, $d\boldsymbol{x} = \nabla_X \boldsymbol{x} \cdot d\boldsymbol{X} = \boldsymbol{F} \cdot d\boldsymbol{X}$. Taking the difference in the squared magnitudes of these elements yields

$$\begin{aligned} d\boldsymbol{x} \cdot d\boldsymbol{x} - d\boldsymbol{X} \cdot d\boldsymbol{X} &= (\nabla_X \boldsymbol{x} \cdot d\boldsymbol{X}) \cdot (\nabla_X \boldsymbol{x} \cdot d\boldsymbol{X}) - d\boldsymbol{X} \cdot d\boldsymbol{X} \\ &= d\boldsymbol{X} \cdot (\boldsymbol{F}^T \cdot \boldsymbol{F} - \mathbf{1}) \cdot d\boldsymbol{X} \stackrel{\text{def}}{=} 2 d\boldsymbol{X} \cdot \boldsymbol{E} \cdot d\boldsymbol{X}. \end{aligned} \tag{B.5}$$

Equation (B.5) defines the so-called *Lagrangian* strain tensor

$$\boldsymbol{E} \stackrel{\text{def}}{=} \tfrac{1}{2}(\boldsymbol{F}^T \cdot \boldsymbol{F} - \mathbf{1}) = \tfrac{1}{2}[\nabla_X \boldsymbol{u} + (\nabla_X \boldsymbol{u})^T + (\nabla_X \boldsymbol{u})^T \cdot \nabla_X \boldsymbol{u}]. \tag{B.6}$$

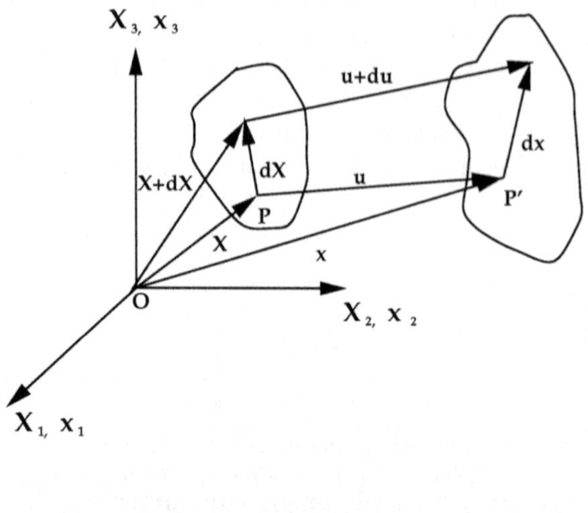

Fig. B.1 Different descriptions of a deforming body

Remark: It should be clear that $d\boldsymbol{x}$ can be reinterpreted as the result of a mapping $\boldsymbol{F} \cdot d\boldsymbol{X} \rightarrow d\boldsymbol{x}$ or a change in configuration (reference to current). One may develop the so-called Eulerian formulations, employing the current configuration coordinates to generate Eulerian strain tensor measures. An important quantity is the Jacobian of the deformation gradient, $J \stackrel{\text{def}}{=} det\boldsymbol{F}$, which relates differential volumes in the reference configuration ($d\omega_0$) to differential volumes in the current configuration ($d\omega$) via $d\omega = J\, d\omega_0$. The Jacobian of the deformation gradient must remain positive, otherwise we obtain physically impossible "negative" volumes. For more details, we refer the reader to the texts of Malvern [1], Gurtin [2], Chandrasekharaiah and Debnath [3].

B.2 Equilibrium/Kinetics of Solid Continua

The balance of linear momentum in the deformed (current) configuration is

$$\underbrace{\int_{\partial\omega} \boldsymbol{t}\, da}_{\text{surface forces}} + \underbrace{\int_{\omega} \rho \boldsymbol{b}\, d\omega}_{\text{body forces}} = \underbrace{\frac{d}{dt}\int_{\omega} \rho \dot{\boldsymbol{u}}\, d\omega}_{\text{inertial forces}}, \tag{B.7}$$

where $\omega \subset \Omega$ is an arbitrary portion of the continuum, with boundary $\partial\omega$, ρ is the material density, $\boldsymbol{b}$ is the body force per unit mass, and $\dot{\boldsymbol{u}}$ is the time derivative of the displacement. The force densities, $\boldsymbol{t}$, are commonly referred to as "surface forces" or tractions.

B.2.1 Postulates on Volume and Surface Quantities

Now, consider a tetrahedron in equilibrium, as shown in Fig. B.2, where a balance of forces yields

$$\boldsymbol{t}^{(n)}\Delta A^{(n)} + \boldsymbol{t}^{(-1)}\Delta A^{(1)} + \boldsymbol{t}^{(-2)}\Delta A^{(2)} + \boldsymbol{t}^{(-3)}\Delta A^{(3)} + \rho\boldsymbol{b}\Delta V = \rho\Delta V\ddot{\boldsymbol{u}}\,, \tag{B.8}$$

where $\Delta A^{(n)}$ is the surface area of the face of the tetrahedron with normal $\boldsymbol{n}$, and ΔV is the tetrahedron volume. As the distance (h) between the tetrahedron base (located at (0,0,0)) and the surface center goes to zero ($h \rightarrow 0$), we have $\Delta A^{(n)} \rightarrow 0 \Rightarrow \frac{\Delta V}{\Delta A^{(n)}} \rightarrow 0$. Geometrically, we have $\frac{\Delta A^{(i)}}{\Delta A^{(n)}} = cos(x_i, x_n) \stackrel{\text{def}}{=} n_i$, and therefore $\boldsymbol{t}^{(n)} + \boldsymbol{t}^{(-1)}cos(x_1, x_n) + \boldsymbol{t}^{(-2)}cos(x_2, x_n) + \boldsymbol{t}^{(-3)}cos(x_3, x_n) = \boldsymbol{0}$. It is clear that forces on the surface areas could be decomposed into three linearly independent components. It is convenient to introduce the concept of stress at a point, representing the surface forces there, pictorially represented by a cube surrounding a point. The fundamental issue that must be resolved is the characterization of these surface forces.

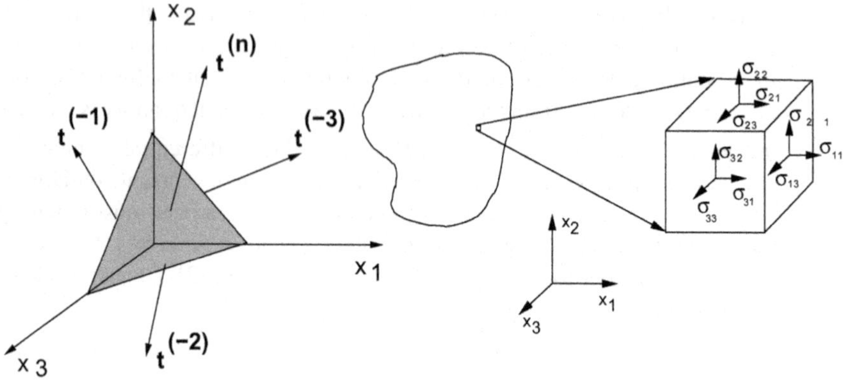

Fig. B.2 Left: Cauchy tetrahedron: a "sectioned point" and Right: Stress at a point

We can represent the surface force density vector, the so-called traction, on a surface by the component representation:

$$\boldsymbol{t}^{(i)} \stackrel{\text{def}}{=} \begin{Bmatrix} \sigma_{i1} \\ \sigma_{i2} \\ \sigma_{i3} \end{Bmatrix}, \tag{B.9}$$

where the second index represents the direction of the component and the first index represents components of the normal to corresponding coordinate plane. Henceforth, we will drop the superscript notation of $\boldsymbol{t}^{(n)}$, where it is implicit that $\boldsymbol{t} \stackrel{\text{def}}{=} \boldsymbol{t}^{(n)} = \boldsymbol{\sigma}^T \cdot \boldsymbol{n}$, where

$$\boldsymbol{\sigma} \stackrel{\text{def}}{=} \begin{bmatrix} \sigma_{11} & \sigma_{12} & \sigma_{13} \\ \sigma_{21} & \sigma_{22} & \sigma_{23} \\ \sigma_{31} & \sigma_{32} & \sigma_{33} \end{bmatrix}, \tag{B.10}$$

or explicitly ($\boldsymbol{t}^{(1)} = -\boldsymbol{t}^{(-1)}, \boldsymbol{t}^{(2)} = -\boldsymbol{t}^{(-2)}, \boldsymbol{t}^{(3)} = -\boldsymbol{t}^{(-3)}$)

$$\boldsymbol{t} = \boldsymbol{t}^{(1)} n_1 + \boldsymbol{t}^{(2)} n_2 + \boldsymbol{t}^{(3)} n_3 = \boldsymbol{\sigma}^T \cdot \boldsymbol{n} = \begin{bmatrix} \sigma_{11} & \sigma_{12} & \sigma_{13} \\ \sigma_{21} & \sigma_{22} & \sigma_{23} \\ \sigma_{31} & \sigma_{32} & \sigma_{33} \end{bmatrix}^T \begin{Bmatrix} n_1 \\ n_2 \\ n_3 \end{Bmatrix}, \tag{B.11}$$

where $\boldsymbol{\sigma}$ is the so-called Cauchy stress tensor.

Remark: In the absence of couple stresses, a balance of angular momentum implies a symmetry of stress, $\boldsymbol{\sigma} = \boldsymbol{\sigma}^T$, and thus the difference in notations becomes immaterial. Explicitly, starting with an angular momentum balance, under the assumptions that no infinitesimal "micro-moments" or so-called couple-stresses exist, then it can be shown that the stress tensor must be symmetric, i.e., $\int_{\partial\omega} \boldsymbol{x} \times \boldsymbol{t}\, da + \int_{\omega} \boldsymbol{x} \times \rho \boldsymbol{b}\, d\omega = \frac{d}{dt} \int_{\omega} \boldsymbol{x} \times \rho \dot{\boldsymbol{u}}\, d\omega$; that is, $\boldsymbol{\sigma}^T = \boldsymbol{\sigma}$. It is somewhat easier to consider a differential element, such as in Fig. B.2, and to simply sum moments about the center. Doing this, one immediately obtains $\sigma_{12} = \sigma_{21}, \sigma_{23} = \sigma_{32}$ and $\sigma_{13} = \sigma_{31}$. Consequently, $\boldsymbol{t} = \boldsymbol{\sigma} \cdot \boldsymbol{n} = \boldsymbol{\sigma}^T \cdot \boldsymbol{n}$.

B.2.2 Balance Law Formulations

Substitution of Eq. B.11 into Eq. B.7 yields ($\omega \subset \Omega$)

$$\underbrace{\int_{\partial\omega} \boldsymbol{\sigma} \cdot \boldsymbol{n}\, da}_{\text{surface forces}} + \underbrace{\int_{\omega} \rho \boldsymbol{b}\, d\omega}_{\text{body forces}} = \underbrace{\frac{d}{dt} \int_{\omega} \rho \dot{\boldsymbol{u}}\, d\omega}_{\text{inertial forces}} . \tag{B.12}$$

A relationship can be determined between the densities in the current and reference configurations, $\int_{\omega} \rho d\omega = \int_{\omega_0} \rho J d\omega_0 = \int_{\omega_0} \rho_0 d\omega_0$. Therefore, the Jacobian can also be interpreted as the ratio of material densities at a point. Since the volume is arbitrary, we can assume that $\rho J = \rho_0$ holds at every point in the body. Therefore, we may write $\frac{d}{dt}(\rho_0) = \frac{d}{dt}(\rho J) = 0$, when the system is mass conservative over time. This leads to writing the last term in Eq. B.12 as $\frac{d}{dt} \int_{\omega} \rho \dot{\boldsymbol{u}}\, d\omega = \int_{\omega_0} \frac{d(\rho J)}{dt} \dot{\boldsymbol{u}}\, d\omega_0 + \int_{\omega_0} \rho \ddot{\boldsymbol{u}} J\, d\omega_0 = \int_{\omega} \rho \ddot{\boldsymbol{u}}\, d\omega$. From Gauss's divergence theorem, and an implicit assumption that $\boldsymbol{\sigma}$ is differentiable, we have $\int_{\omega} (\nabla_x \cdot \boldsymbol{\sigma} + \rho \boldsymbol{b} - \rho \ddot{\boldsymbol{u}})\, d\omega = \boldsymbol{0}$. If the volume is argued as being arbitrary, then the integrand must be equal to zero at every point, yielding

$$\nabla_x \cdot \boldsymbol{\sigma} + \rho \boldsymbol{b} = \rho \ddot{\boldsymbol{u}} . \tag{B.13}$$

B.3 Referential Descriptions of Balance Laws and Nanson's Formula

Although we will not consider finite deformation problems in this monograph, some important concepts will be useful later in the context of mapping from one configuration to the next. In many cases it is quite difficult to perform a stress analysis, for finite deformation solid mechanics problems, in the current configuration, primarily because it is unknown a priori. Therefore all quantities are usually transformed ("pulled") back to the original coordinates, the referential frame. Therefore, it is preferable to think of a formulation in terms of the referential fixed coordinated $\boldsymbol{X}$, a so-called *Lagrangian* formulation. With this in mind there are two commonly used referential measures of stresses. We start by a purely mathematical result, leading to the so-called Nanson formula for transformation of surface elements. Consider the cross product of two differential line elements in a current configuration, $d\boldsymbol{x}^{(1)} \times d\boldsymbol{x}^{(2)} = (\boldsymbol{F} \cdot d\boldsymbol{X}^{(1)}) \times (\boldsymbol{F} \cdot d\boldsymbol{X}^{(2)})$. An important vector identity (see Chandriashakiah and Debnath [3]) for a tensor $\boldsymbol{T}$ and two first-order vectors $\boldsymbol{a}$ and $\boldsymbol{b}$ is $(\boldsymbol{T} \cdot \boldsymbol{a}) \times (\boldsymbol{T} \cdot \boldsymbol{b}) = \boldsymbol{T}^* \cdot (\boldsymbol{a} \times \boldsymbol{b})$, where the $\boldsymbol{T}^*$ is the transpose of the adjoint defined by $\boldsymbol{T}^* \stackrel{\text{def}}{=} (det\boldsymbol{T})\boldsymbol{T}^{-T}$. This leads to $(det\boldsymbol{T})\boldsymbol{1} = \boldsymbol{T}^T \cdot \boldsymbol{T}^*$. Applying the result we have $d\boldsymbol{x}^{(1)} \times d\boldsymbol{x}^{(2)} = \boldsymbol{F}^* \cdot (d\boldsymbol{X}^{(1)} \times d\boldsymbol{X}^{(2)})$ and $\boldsymbol{F}^T \cdot (d\boldsymbol{x}^{(1)} \times d\boldsymbol{x}^{(2)}) = (det\boldsymbol{F})\boldsymbol{1} \cdot (d\boldsymbol{X}^{(1)} \times d\boldsymbol{X}^{(2)})$. This leads to $\boldsymbol{F}^T \cdot \boldsymbol{n} da = (det\boldsymbol{F})\boldsymbol{n}_0 da_0$. This is the so-called Nanson formula. Knowing this, we now formulate the equations of equilibrium in the current or reference configuration (Fig. B.3).

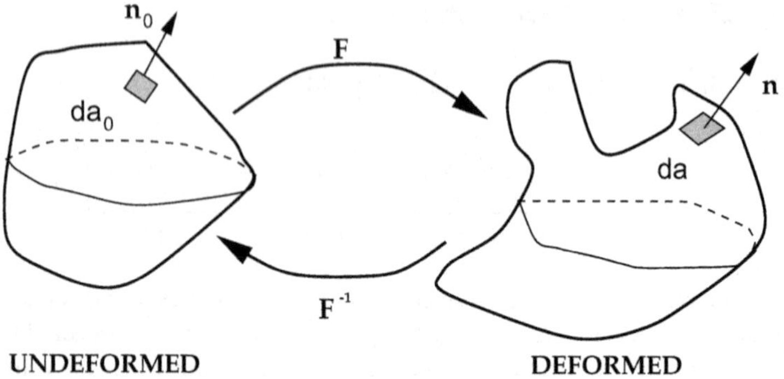

Fig. B.3 A current and reference surface element

Consider two surface elements: one on the current configuration and one on a reference configuration. If we form a new kind of stress tensor, call it $\boldsymbol{P}$, such that the amount of force is the same we have $\boldsymbol{P}\cdot\boldsymbol{n}_0 da_0 = \boldsymbol{\sigma}\cdot\boldsymbol{n} da = \boldsymbol{\sigma}\cdot\boldsymbol{F}^{-T}(det\boldsymbol{F})\cdot\boldsymbol{n}_0 da_0$ which implies $\boldsymbol{P} = \boldsymbol{\sigma}\cdot\boldsymbol{F}^{-T}(det\boldsymbol{F})$. The tensor $\boldsymbol{P}$ is called the first Piola–Kirchhoff stress and gives the actual force on the current area, but calculated per unit area of reference area. However, it is not symmetric, and this sometimes causes difficulties in an analysis. Therefore, we symmetrize it by $\boldsymbol{F}^{-1}\cdot\boldsymbol{P} = \boldsymbol{S} = \boldsymbol{S}^T = \boldsymbol{F}^{-1}\cdot\boldsymbol{\sigma}\cdot\boldsymbol{F}^{-T}(det\boldsymbol{F})$. The tensor $\boldsymbol{S}$ is called the second Piola–Kirchhoff stress. By definition we have $\int_{\partial\omega_0}\boldsymbol{n}_0\cdot\boldsymbol{P}\,da_0 = \int_{\partial\omega}\boldsymbol{n}\cdot\boldsymbol{\sigma}\,da$, and thus

$$\underbrace{\int_{\partial\omega_0}\boldsymbol{n}_0\cdot\boldsymbol{P}\,da_0}_{\text{surface forces}}+\underbrace{\int_{\omega_0}\boldsymbol{f}J\,d\omega}_{\text{body forces}}=\int_{\omega_0}\rho_0\frac{d\dot{\boldsymbol{u}}}{dt}\,d\omega_0, \tag{B.14}$$

and therefore

$$\underbrace{\int_{\omega_0}\nabla_X\cdot\boldsymbol{P}\,d\omega_0}_{\text{surface forces}}+\underbrace{\int_{\omega_0}\boldsymbol{f}J\,d\omega_0}_{\text{body forces}}=\int_{\omega_0}\rho_0\frac{d\dot{\boldsymbol{u}}}{dt}\,d\omega_0. \tag{B.15}$$

Since $\boldsymbol{P} = \boldsymbol{F}\cdot\boldsymbol{S}$, $\int_{\omega_0}\nabla_X\cdot(\boldsymbol{F}\cdot\boldsymbol{S})\,d\omega_0 + \int_{\omega_0}\boldsymbol{f}J\,d\omega_0 = \int_{\omega_0}\rho_0\frac{d\dot{\boldsymbol{u}}}{dt}\,d\omega_0$. Since the control volume is arbitrary, we have

$$\nabla_X\cdot\boldsymbol{P}+\boldsymbol{f}J=\rho_0\frac{d\dot{\boldsymbol{u}}}{dt} \quad \text{or} \quad \nabla_X\cdot(\boldsymbol{F}\cdot\boldsymbol{S})+\boldsymbol{f}J=\rho_0\frac{d\dot{\boldsymbol{u}}}{dt}. \tag{B.16}$$

B.4 The First Law of Thermodynamics/An Energy Balance

The interconversions of mechanical, thermal, and chemical energy in a system are governed by the first law of thermodynamics, which states that the time rate of change of the total energy, $\mathcal{K}+\mathcal{I}$, is equal to the mechanical power, $\mathcal{P}$, and the net heat supplied, $\mathcal{H}+\mathcal{Q}$, i.e., $\frac{d}{dt}(\mathcal{K}+\mathcal{I}) = \mathcal{P}+\mathcal{H}+\mathcal{Q}$. Here the kinetic energy of a subvolume of material contained in Ω, denoted ω, is $\mathcal{K} \stackrel{\text{def}}{=} \int_{\omega} \frac{1}{2}\rho\dot{\boldsymbol{u}} \cdot \dot{\boldsymbol{u}}\, d\omega$; the power (rate of work) of the external forces acting on ω is given by $\mathcal{P} \stackrel{\text{def}}{=} \int_{\omega} \rho \boldsymbol{b} \cdot \dot{\boldsymbol{u}}\, d\omega + \int_{\partial\omega} \boldsymbol{\sigma} \cdot \boldsymbol{n} \cdot \dot{\boldsymbol{u}}\, da$; the heat flow into the volume by conduction is $\mathcal{Q} \stackrel{\text{def}}{=} -\int_{\partial\omega} \boldsymbol{q} \cdot \boldsymbol{n}\, da = -\int_{\omega} \nabla_x \cdot \boldsymbol{q}\, d\omega$, $\boldsymbol{q}$ being the heat flux; the heat generated due to sources, *such as chemical reactions*, is $\mathcal{H} \stackrel{\text{def}}{=} \int_{\omega} \rho z\, d\omega$, where z is the reaction source rate per unit mass; and the internal energy is $\mathcal{I} \stackrel{\text{def}}{=} \int_{\omega} \rho \mathrm{w}\, d\omega$, w being the internal energy per unit mass. Differentiating the kinetic energy yields

$$\begin{aligned}\frac{d\mathcal{K}}{dt} = \frac{d}{dt}\int_{\omega} \frac{1}{2}\rho\dot{\boldsymbol{u}} \cdot \dot{\boldsymbol{u}}\, d\omega &= \int_{\omega_0} \frac{d}{dt}\frac{1}{2}(\rho J \dot{\boldsymbol{u}} \cdot \dot{\boldsymbol{u}})\, d\omega_0 \\ &= \int_{\omega_0} (\frac{d}{dt}\rho_0)\frac{1}{2}\dot{\boldsymbol{u}} \cdot \dot{\boldsymbol{u}}\, d\omega_0 + \int_{\omega} \rho \frac{d}{dt}\frac{1}{2}(\dot{\boldsymbol{u}} \cdot \dot{\boldsymbol{u}})\, d\omega \\ &= \int_{\omega} \rho\dot{\boldsymbol{u}} \cdot \ddot{\boldsymbol{u}}\, d\omega,\end{aligned} \tag{B.17}$$

where we have assumed that the mass in the system is constant. We also have

$$\frac{d\mathcal{I}}{dt} = \frac{d}{dt}\int_{\omega} \rho \mathrm{w}\, d\omega = \frac{d}{dt}\int_{\omega_0} \rho J \mathrm{w}\, d\omega_0 = \int_{\omega_0} \underbrace{\frac{d}{dt}(\rho_0)}_{=0} \mathrm{w}\, d\omega_0 + \int_{\omega} \rho\dot{\mathrm{w}}\, d\omega = \int_{\omega} \rho\dot{\mathrm{w}}\, d\omega. \tag{B.18}$$

By using the divergence theorem, we obtain

$$\int_{\partial\omega} \boldsymbol{\sigma} \cdot \boldsymbol{n} \cdot \dot{\boldsymbol{u}}\, da = \int_{\omega} \nabla_x \cdot (\boldsymbol{\sigma} \cdot \dot{\boldsymbol{u}})\, d\omega = \int_{\omega} (\nabla_x \cdot \boldsymbol{\sigma}) \cdot \dot{\boldsymbol{u}}\, d\omega + \int_{\omega} \boldsymbol{\sigma} : \nabla_x \dot{\boldsymbol{u}}\, d\omega. \tag{B.19}$$

Combining the results, and enforcing a balance of linear momentum, leads to

$$\begin{aligned}&\int_{\omega} (\rho\dot{\mathrm{w}} + \dot{\boldsymbol{u}} \cdot (\rho\ddot{\boldsymbol{u}} - \nabla_x \cdot \boldsymbol{\sigma} - \rho\boldsymbol{b}) - \boldsymbol{\sigma} : \nabla_x \dot{\boldsymbol{u}} + \nabla_x \cdot \boldsymbol{q} - \rho z)\, d\omega = \\ &\int_{\omega} (\rho\dot{\mathrm{w}} - \boldsymbol{\sigma} : \nabla_x \dot{\boldsymbol{u}} + \nabla_x \cdot \boldsymbol{q} - \rho z)\, d\omega = 0.\end{aligned} \tag{B.20}$$

Since the volume ω is arbitrary, the integrand must hold locally and we have

$$\rho\dot{\mathrm{w}} - \boldsymbol{\sigma} : \nabla_x \dot{\boldsymbol{u}} + \nabla_x \cdot \boldsymbol{q} - \rho z = 0. \tag{B.21}$$

When dealing with multifield problems, this equation is used extensively.

B.5 Linearly Elastic Constitutive Equations

We now discuss relationships between the stress and strain, so-called *material laws* or *constitutive relations* for linearly elastic cases (infinitesimal deformations).

B.5.1 The Infinitesimal Strain Case

In infinitesimal deformation theory, the displacement gradient components are considered small enough that higher-order terms such as $(\nabla_X \boldsymbol{u})^T \cdot \nabla_X \boldsymbol{u}$ and $(\nabla_x \boldsymbol{u})^T \cdot \nabla_x \boldsymbol{u}$ can be neglected in the strain measure $\boldsymbol{E} = \frac{1}{2}(\nabla_X \boldsymbol{u} + (\nabla_X \boldsymbol{u})^T + (\nabla_X \boldsymbol{u})^T \cdot \nabla_X \boldsymbol{u})$, leading to $\boldsymbol{E} \approx \boldsymbol{\epsilon} \stackrel{\text{def}}{=} \frac{1}{2}[\nabla_X \boldsymbol{u} + (\nabla_X \boldsymbol{u})^T]$. If the displacement gradients are small compared with unity, $\boldsymbol{\epsilon}$ coincides closely with $\boldsymbol{E}$. If we assume that $\frac{\partial}{\partial \boldsymbol{X}} \approx \frac{\partial}{\partial \boldsymbol{x}}$, we may use $\boldsymbol{E}$ or $\boldsymbol{\epsilon}$ interchangeably. Usually $\boldsymbol{\epsilon}$ is the symbol used for infinitesimal strains. Furthermore, to avoid confusion, when using models employing the geometrically linear infinitesimal strain assumption we use the symbol of ∇ with no $\boldsymbol{X}$ or $\boldsymbol{x}$ subscript. Hence, the infinitesimal strains are defined by

$$\boldsymbol{\epsilon} = \frac{1}{2}(\nabla \boldsymbol{u} + (\nabla \boldsymbol{u})^T). \tag{B.22}$$

B.5.2 Linear Elastic Constitutive Laws

If we neglect thermal effects, Eq. B.21 implies $\rho \dot{\text{w}} = \boldsymbol{\sigma} : \nabla_x \dot{\boldsymbol{u}}$ which, in the infinitesimal strain linearly elastic case, is $\rho \dot{\text{w}} = \boldsymbol{\sigma} : \dot{\boldsymbol{\epsilon}}$. From the chain rule of differentiation we have

$$\rho \dot{\text{w}} = \rho \frac{\partial \text{w}}{\partial \boldsymbol{\epsilon}} : \frac{d\boldsymbol{\epsilon}}{dt} = \boldsymbol{\sigma} : \dot{\boldsymbol{\epsilon}} \Rightarrow \boldsymbol{\sigma} = \rho \frac{\partial \text{w}}{\partial \boldsymbol{\epsilon}}. \tag{B.23}$$

The starting point to develop a constitutive theory is to assume a stored elastic energy function exists, a function denoted $W \stackrel{\text{def}}{=} \rho \text{w}$, which depends only on the mechanical deformation. The simplest function that fulfills $\boldsymbol{\sigma} = \rho \frac{\partial \text{w}}{\partial \boldsymbol{\epsilon}}$ is $W = \frac{1}{2}\boldsymbol{\epsilon} : \mathbb{E} : \boldsymbol{\epsilon}$, where $\mathbb{E}$ is the fourth rank elasticity tensor. Such a function satisfies the intuitive physical requirement that, for any small strain from an undeformed state, energy must be stored in the material. Alternatively, a small strain material law can be derived from $\boldsymbol{\sigma} = \frac{\partial W}{\partial \boldsymbol{\epsilon}}$ and $W \approx c_0 + \mathbf{c}_1 : \boldsymbol{\epsilon} + \frac{1}{2}\boldsymbol{\epsilon} : \mathbb{E} : \boldsymbol{\epsilon} + \ldots$ which implies $\boldsymbol{\sigma} \approx \mathbf{c}_1 + \mathbb{E} : \boldsymbol{\epsilon} + \ldots$. We are free to set $c_0 = 0$ (it is arbitrary) in order to have zero strain energy at zero strain, and furthermore, we assume that no stresses exist in the reference state ($\mathbf{c}_1 = \mathbf{0}$). With these assumptions, we obtain the familiar relation

$$\boldsymbol{\sigma} = \mathbb{E} : \boldsymbol{\epsilon}. \tag{B.24}$$

This is a linear relation between stresses and strains. The existence of a strictly positive stored energy function in the reference configuration implies that the linear elasticity tensor must have positive eigenvalues at every point in the body. Typically, different materials are classified according to the number of independent components in $\mathbb{E}$. In theory, $\mathbb{E}$ has 81 components, since it is a fourth-order tensor relating 9 components of stress to strain. However, the number of components can be reduced to 36 since the stress and strain tensors are symmetric. This is observed from the matrix representation[2] of $\mathbb{E}$:

$$\underbrace{\begin{Bmatrix} \sigma_{11} \\ \sigma_{22} \\ \sigma_{33} \\ \sigma_{12} \\ \sigma_{23} \\ \sigma_{31} \end{Bmatrix}}_{\stackrel{\text{def}}{=}\{\sigma\}} = \underbrace{\begin{bmatrix} E_{1111} & E_{1122} & E_{1133} & E_{1112} & E_{1123} & E_{1113} \\ E_{2211} & E_{2222} & E_{2233} & E_{2212} & E_{2223} & E_{2213} \\ E_{3311} & E_{3322} & E_{3333} & E_{3312} & E_{3323} & E_{3313} \\ E_{1211} & E_{1222} & E_{1233} & E_{1212} & E_{1223} & E_{1213} \\ E_{2311} & E_{2322} & E_{2333} & E_{2312} & E_{2323} & E_{2313} \\ E_{1311} & E_{1322} & E_{1333} & E_{1312} & E_{1323} & E_{1313} \end{bmatrix}}_{\stackrel{\text{def}}{=}[\mathbb{E}]} \underbrace{\begin{Bmatrix} \epsilon_{11} \\ \epsilon_{22} \\ \epsilon_{33} \\ 2\epsilon_{12} \\ 2\epsilon_{23} \\ 2\epsilon_{31} \end{Bmatrix}}_{\stackrel{\text{def}}{=}\{\epsilon\}}. \tag{B.25}$$

The existence of a scalar energy function forces $\mathbb{E}$ to be symmetric since the strains are symmetric, in other words $W = \frac{1}{2}\epsilon : \mathbb{E} : \epsilon = \frac{1}{2}(\epsilon : \mathbb{E} : \epsilon)^T = \frac{1}{2}\epsilon^T : \mathbb{E}^T : \epsilon^T = \frac{1}{2}\epsilon : \mathbb{E}^T : \epsilon$ which implies $\mathbb{E}^T = \mathbb{E}$. Consequently, $\mathbb{E}$ has only 21 independent components. The nonnegativity of W imposes the restriction that $\mathbb{E}$ remains positive definite. At this point, based on many factors that depend on the material microstructure, it can be shown that the components of $\mathbb{E}$ may be written in terms of anywhere between 21 and 2 independent parameters. Accordingly, for isotropic materials, we have two planes of symmetry and an infinite number of planes of directional independence (two free components), yielding

$$\mathbb{E} \stackrel{\text{def}}{=} \begin{bmatrix} \kappa + \frac{4}{3}\mu & \kappa - \frac{2}{3}\mu & \kappa - \frac{2}{3}\mu & 0 & 0 & 0 \\ \kappa - \frac{2}{3}\mu & \kappa + \frac{4}{3}\mu & \kappa - \frac{2}{3}\mu & 0 & 0 & 0 \\ \kappa - \frac{2}{3}\mu & \kappa - \frac{2}{3}\mu & \kappa + \frac{4}{3}\mu & 0 & 0 & 0 \\ 0 & 0 & 0 & \mu & 0 & 0 \\ 0 & 0 & 0 & 0 & \mu & 0 \\ 0 & 0 & 0 & 0 & 0 & \mu \end{bmatrix}. \tag{B.26}$$

In this case, we have

$$\mathbb{E} : \epsilon = 3\kappa \frac{tr\epsilon}{3}\mathbf{1} + 2\mu\epsilon' \Rightarrow \epsilon : \mathbb{E} : \epsilon = 9\kappa(\frac{tr\epsilon}{3})^2 + 2\mu\epsilon' : \epsilon', \tag{B.27}$$

where $tr\epsilon = \epsilon_{ii}$ and $\epsilon' = \epsilon - \frac{1}{3}(tr\epsilon)\mathbf{1}$ is the deviatoric strain. The eigenvalues of an isotropic elasticity tensor are $(3\kappa, 2\mu, 2\mu, \mu, \mu, \mu)$. Therefore, we must have $\kappa > 0$ and $\mu > 0$ to retain positive definiteness of $\mathbb{E}$. All of the material components of $\mathbb{E}$ may be spatially variable, as in the case of composite media.

[2]The symbol $[\cdot]$ is used to indicate the matrix notation equivalent to a tensor form, while $\{\cdot\}$ is used to indicate the vector representation.

B.5.3 Material Component Interpretation

There are a variety of ways to write isotropic constitutive laws, each time with a physically meaningful pair of material values.

Splitting the strain

It is sometimes important to split infinitesimal strains into two physically meaningful parts

$$\epsilon = \frac{tr\epsilon}{3}\mathbf{1} + (\epsilon - \frac{tr\epsilon}{3}\mathbf{1}). \tag{B.28}$$

An expansion of the Jacobian of the deformation gradient yields $J = det(\mathbf{1} + \nabla_X \boldsymbol{u}) \approx 1 + tr\nabla_X \boldsymbol{u} + \mathcal{O}(\nabla_X \boldsymbol{u}) = 1 + tr\epsilon + ...$ Therefore, with infinitesimal strains, $(1+tr\epsilon)d\omega_0 = d\omega$ and we can write $tr\epsilon = \frac{d\omega - d\omega_0}{d\omega_0}$. Hence, $tr\epsilon$ is associated with the *volumetric part of the deformation*. Furthermore, since $\epsilon' \stackrel{\text{def}}{=} \epsilon - \frac{tr\epsilon}{3}\mathbf{1}$, the so-called strain deviator describes distortion in the material.

Infinitesimal strain material laws

The stress $\boldsymbol{\sigma}$ can be split into two parts (dilatational and a deviatoric):

$$\boldsymbol{\sigma} = \frac{tr\boldsymbol{\sigma}}{3}\mathbf{1} + (\boldsymbol{\sigma} - \frac{tr\boldsymbol{\sigma}}{3}\mathbf{1}) \stackrel{\text{def}}{=} -p\mathbf{1} + \boldsymbol{\sigma}', \tag{B.29}$$

where we call the symbol p the hydrostatic pressure and $\boldsymbol{\sigma}'$ the stress deviator. With (B.27) we write

$$p = -3\,\kappa \left(\frac{tr\epsilon}{3}\right) \quad \text{and} \quad \boldsymbol{\sigma}' = 2\,\mu\,\epsilon'. \tag{B.30}$$

This is one form of Hooke's Law. The resistance to change in the volume is measured by κ. We note that $(\frac{tr\boldsymbol{\sigma}}{3}\mathbf{1})' = \mathbf{0}$, which indicates that this part of the stress produces no distortion.

Another fundamental form of Hooke's law is

$$\boldsymbol{\sigma} = \frac{E}{1+\nu}\left(\epsilon + \frac{\nu}{1-2\nu}(tr\epsilon)\mathbf{1}\right), \tag{B.31}$$

and the inverse form

$$\epsilon = \frac{1+\nu}{E}\boldsymbol{\sigma} - \frac{\nu}{E}(tr\boldsymbol{\sigma})\mathbf{1}. \tag{B.32}$$

To interpret the material values, consider an idealized uniaxial tension test (pulled in the x_1 direction inducing a uniform stress state) where $\sigma_{12} = \sigma_{13} = \sigma_{23} = 0$, which implies $\epsilon_{12} = \epsilon_{13} = \epsilon_{23} = 0$. Also, we have $\sigma_{22} = \sigma_{33} = 0$. Under these conditions we have $\sigma_{11} = E\epsilon_{11}$ and $\epsilon_{22} = \epsilon_{33} = -\nu\epsilon_{11}$. Therefore, E, Young's modulus, is the ratio of the uniaxial stress to the corresponding strain component. The Poisson ratio, ν, is the ratio of the transverse strains to the uniaxial strain.

Another commonly used set of stress–strain forms are the Lamé relations:

$$\boldsymbol{\sigma} = \lambda(tr\boldsymbol{\epsilon})\mathbf{1} + 2\mu\boldsymbol{\epsilon} \quad \text{or} \quad \boldsymbol{\epsilon} = -\frac{\lambda}{2\mu(3\lambda + 2\mu)}(tr\boldsymbol{\sigma}\mathbf{1}) + \frac{\boldsymbol{\sigma}}{2\mu}. \tag{B.33}$$

To interpret the material values, consider a homogeneous pressure test (uniform stress), where $\sigma_{12} = \sigma_{13} = \sigma_{23} = 0$, and where $\sigma_{11} = \sigma_{22} = \sigma_{33}$. Under these conditions, we have

$$\kappa = \lambda + \frac{2}{3}\mu = \frac{E}{3(1-2\nu)} \quad \text{and} \quad \mu = \frac{E}{2(1+\nu)}, \tag{B.34}$$

and consequently

$$\frac{\kappa}{\mu} = \frac{2(1+\nu)}{3(1-2\nu)}. \tag{B.35}$$

We observe that $\frac{\kappa}{\mu} \to \infty$ implies $\nu \to \frac{1}{2}$, and $\frac{\kappa}{\mu} \to 0$ implies $\Rightarrow \nu \to -1$. Therefore, since both κ and μ must be positive and finite, this implies $-1 < \nu < 1/2$ and $0 < E < \infty$. For example, some polymeric foams exhibit $\nu < 0$, steels $\nu \approx 0.3$, and some forms of rubber have $\nu \to 1/2$. We note that λ *can be positive or negative.* For more details, see Malvern [1], Gurtin [2], Chandrasekharaiah and Debnath [3].

B.6 Related Physical Concepts

In closing, we briefly consider two other commonly encountered physical scenarios which are formally related to mechanical equilibrium.

B.6.1 Heat Conduction

We recall from our thermodynamic analysis the first law in the current configuration

$$\rho\dot{w} - \boldsymbol{\sigma} : \nabla_x \dot{\boldsymbol{u}} + \nabla_x \cdot \boldsymbol{q} - \rho z = 0, \tag{B.36}$$

or in the reference configuration as

$$\rho_0\dot{w} - \boldsymbol{S} : \dot{\boldsymbol{E}} + \nabla_X \cdot \boldsymbol{q}_0 - \rho_0 z = 0, \tag{B.37}$$

where $\boldsymbol{q}_0 = \boldsymbol{q}J \cdot \boldsymbol{F}^{-T}$. When (1) the deformations are ignored, $\boldsymbol{u} = \mathbf{0}$, thus $\boldsymbol{S} : \dot{\boldsymbol{E}} = 0$, (2) the stored energy is purely thermal, described by $\rho_0\dot{w} = \rho_0 C\dot{\theta}$, where C is the heat capacity, (3) the reactions are zero, $\rho_0 z = 0$, (4) the variation in time is ignored, i.e., steady-state, and (5) $\boldsymbol{q}_0 = -\mathbb{K} \cdot \nabla\theta$, where the conductivity tensor

$I\!K \in I\!R^{3\times3}$ is a spatially varying symmetric bounded positive definite tensor-valued function, then we arrive at the familiar equation of linear heat conduction:

$$\nabla_X \cdot (I\!K \cdot \nabla_X \theta) = \rho_0 C\dot{\theta}. \tag{B.38}$$

If the variation in time is ignored, i.e., steady-state conditions are enforced:

$$\nabla_X \cdot (I\!K \cdot \nabla_X \theta) = 0. \tag{B.39}$$

B.6.2 Solid-State Diffusion-Reaction

Consider a structure which occupies an open bounded domain in $\Omega \in I\!R^3$, with boundary $\partial\Omega$. The boundary consists of Γ_c and Γ_g, where the solute concentrations (c) and solute fluxes are, respectively, specified. The diffusive properties of the heterogeneous material are characterized by a spatially varying diffusivity $I\!D_0 \in I\!R^{3\times3}$, which is assumed to be a symmetric bounded positive definite tensor-valued function. The mass balance for a small diffusing species, denoted by the normalized concentration of the solute c (molecules per unit volume), in an arbitrary subvolume of material contained within Ω, denoted ω, consists of a storage term ($\dot{c}$), a reaction term ($\dot{s}$), and an inward normal flux term ($-\boldsymbol{G} \cdot \boldsymbol{n}$), leading to $\int_\omega (\dot{c} + \dot{s})\, d\omega = -\int_{\partial\omega} \boldsymbol{G} \cdot \boldsymbol{n}\, da$. It is a classical *stoichiometrically inexact* approximation to assume that the diffusing species reacts (is created or destroyed) in a manner such that the rate of production of the reactant (s) is directly proportional to the concentration of the diffusing species itself and the rate of change of the diffusing species, $\dot{s} = \tau c + \varpi\dot{c}$. Here, $\tau = \tau_0 e^{-\frac{Q}{R\theta}}$ and $\varpi = \varpi_0 e^{-\frac{\mathcal{Q}}{R\theta}}$, where τ_0 and ϖ_0 are rate constants, Q and $\mathcal{Q}$ ($Q \neq \mathcal{Q}$) are activation energies per mole of diffusive species, R is the universal gas constant, and θ is the temperature. Upon substitution of these relations into the conservation law for the diffusing species, and after using the divergence theorem, since the volume ω is arbitrary, one has a Fickian diffusion-reaction model in strong form, assuming $\boldsymbol{G} = -I\!D \cdot \nabla_x c$

$$\dot{c} = \nabla_x \cdot (I\!D \cdot \nabla_x c) - \tau c - \varpi\dot{c} \Rightarrow \dot{c}(1+\varpi) = \nabla_x \cdot (I\!D \cdot \nabla_x c) - \tau c. \tag{B.40}$$

When $\tau_0 > 0$, the diffusing species is destroyed as it reacts, while $\tau_0 < 0$ means that the diffusing species is created as it reacts, i.e., an autocatalytic or "chain" reaction occurs. We will only consider the nonautocatalytic case in this work. Also, depending on the sign of ϖ_0, effectively the process will have an accelerated or decelerated diffusivity as well as accelerated or decelerated reactivity. In Eq. E.14, $I\!D$ is the diffusivity tensor (area per unit time). If we ignore reactions and time dependency, and assume that the domain is not deforming, we then arrive at the familiar

$$\nabla_X \cdot (I\!D \cdot \nabla_X c) = 0. \tag{B.41}$$

B.6.3 Conservation Law Families

In summary we have the following related linearized steady-state forms (with no body forces in mechanical equilibrium)

$$\boxed{\begin{aligned} &\nabla_X \cdot (\mathbb{E} : \nabla_X \boldsymbol{u}) = \boldsymbol{0}, \\ &\nabla_X \cdot (\mathbb{K} \cdot \nabla_X \theta) = 0, \\ &\nabla_X \cdot (\mathbb{D} \cdot \nabla_X c) = 0, \end{aligned}} \tag{B.42}$$

which stem from the following coupled, time-transient, nonlinear equations:

$$\boxed{\begin{aligned} &\nabla_x \cdot \boldsymbol{\sigma} + \rho \boldsymbol{b} = \rho \ddot{u} \\ &\nabla_x \cdot \boldsymbol{q} - \boldsymbol{\sigma} : \nabla_x \dot{\boldsymbol{u}} - \rho z = -\rho \dot{w}, \\ &\nabla_x \cdot \boldsymbol{G} + \tau c + \varpi \dot{c} = -\dot{c}. \end{aligned}} \tag{B.43}$$

From this point forth, we consider infinitesimal deformations, and we drop the explicit reference to differentiation with respect to X or x, under the assumption that they are one and the same at infinitesimal deformations

$$\boxed{\begin{aligned} &\nabla \cdot (\mathbb{E} : \nabla \boldsymbol{u}) = \boldsymbol{0}, \\ &\nabla \cdot (\mathbb{K} \cdot \nabla \theta) = 0, \\ &\nabla \cdot (\mathbb{D} \cdot \nabla c) = 0. \end{aligned}} \tag{B.44}$$

Furthermore, we shall use the notation $\boldsymbol{x}$ to indicate the location of a point in space, under the assumption of infinitesimal deformations, where the difference between X and x is considered insignificant.

Appendix C Convergence of Recursive Iterative Schemes

C

Recursive iterative schemes arise frequently in computational mechanics, for example, in implicit time-stepping, domain decomposition, etc. To understand the convergence of such iterative schemes, consider a general system of coupled equations given by

$$\mathcal{A}(s) = \mathcal{F}, \tag{C.1}$$

where s is a solution, and where it is assumed that the operator, $\mathcal{A}$, is invertible. One desires that the sequence of iterated solutions, s^I, $I = 1, 2, ...,$ converge to $\mathcal{A}^{-1}(\mathcal{F})$ as $I \rightarrow \infty$. It is assumed that the Ith iterate can be represented by some arbitrary function $s^I = \mathcal{T}^I(\mathcal{A}, \mathcal{F}, s^{I-1})$. One makes the following split

$$s^I = \mathcal{G}^I(s^{I-1}) + r^I. \tag{C.2}$$

For this method to be useful the exact solution should be reproduced. In other words, when $s = \mathcal{A}^{-1}(\mathcal{F})$, then

$$s = \mathcal{A}^{-1}(\mathcal{F}) = \mathcal{G}^I(\mathcal{A}^{-1}(\mathcal{F})) + r^I. \tag{C.3}$$

Therefore, one has the following consistency condition

$$r^I = \mathcal{A}^{-1}(\mathcal{F}) - \mathcal{G}^I(\mathcal{A}^{-1}(\mathcal{F})), \tag{C.4}$$

and as a consequence,

$$s^I = \mathcal{G}^I(s^{I-1}) + \mathcal{A}^{-1}(\mathcal{F}) - \mathcal{G}^I(\mathcal{A}^{-1}(\mathcal{F})). \tag{C.5}$$

Convergence of the iteration can be studied by defining the error vector:

$$\begin{aligned} e^I &= s^I - s = s^I - \mathcal{A}^{-1}(\mathcal{F}) \\ &= \mathcal{G}^I(s^{I-1}) + \mathcal{A}^{-1}(\mathcal{F}) - \mathcal{G}^I(\mathcal{A}^{-1}(\mathcal{F})) - \mathcal{A}^{-1}(\mathcal{F}) \\ &= \mathcal{G}^I(s^{I-1}) - \mathcal{G}^I(\mathcal{A}^{-1}(\mathcal{F})). \end{aligned} \tag{C.6}$$

T. I. Zohdi, *A Finite Element Primer for Beginners*, The Basics,
https://doi.org/10.1007/978-3-319-70428-9

One sees that, if $\mathcal{G}^I$ is linear and invertible, the above reduces to

$$\boldsymbol{e}^I = \mathcal{G}^I(\boldsymbol{s}^{I-1} - \mathcal{A}^{-1}(\mathcal{F})) = \mathcal{G}^I(\boldsymbol{e}^{I-1}). \tag{C.7}$$

Therefore, if the spectral radius of $\mathcal{G}^I$, i.e., the magnitude of its largest eigenvalue, is less than unity for each iteration I, then $\boldsymbol{e}^I \rightarrow \boldsymbol{0}$ for any arbitrary starting solution $\boldsymbol{s}^{I=0}$ as $I \rightarrow \infty$.

Appendix D
Selected in-Class Exam Problems

D

The problems in this chapter are selected from exams given over the last 15 years UC Berkeley.

D.1 Sample Problem 1

- (a) Concisely explain the classical Galerkin method (without weak form) by considering a simple one-dimensional differential equation, written in the following abstract form

$$A(u) = f, \tag{D.1}$$

where $u(0) = c_1$ and $u(L) = c_2$.
- (b) Analytically solve the following two-point boundary value problem:

$$\frac{d^2u}{dx^2} + bu + sin(ax) = 0 \qquad (BC's: u(0) = 0,\ u(L) = 0), \tag{D.2}$$

with domain size $(0, L)$ and where $a > 0$ and $b > 0$ are constants.
- (c) Using *Galerkin's method* (no weak form), with the approximation

$$u(x) \approx u^{app}(x) = a_1(x - xL) + a_2(x - xL)^2 + a_3 sin(6\pi x/L) + a_4 sin(8\pi x/L) \tag{D.3}$$

generate the system of equations needed to determine a_1 through a_5 to approximately solve Eq. D.2. Set up the integrals. **Put in matrix form, but do not solve.**
- (d) Give expressions for the errors in the H_1, H_2, and H_3 norms, respectively? Which error measure is higher and why?
- (e) List two major difficulties encountered when using the classical Galerkin method for more complex problems, which led us eventually to the finite element method.

T. I. Zohdi, *A Finite Element Primer for Beginners*, The Basics,
https://doi.org/10.1007/978-3-319-70428-9

- (f) Repeat for

$$\frac{d^2u}{dx^2} + au = e^{bx} \qquad (BC's: u(0) = 0,\ u(L) = 0), \tag{D.4}$$

and

$$u(x) \approx u^{app}(x) = a_1(x^4 - xL^3) + a_2(x^3 - xL^2) + a_3(x^2 - xL) + a_4(x - L), \tag{D.5}$$

D.2 Sample Problem 2

If you were given the following

$$E\frac{d^2u}{dx^2} + ku = 0, \tag{D.6}$$

$u(0) = c_1$, $u(L) = c_2$, what is a *quick* way to determine (approximately) the minimum elements would one need to capture the basic physics if E and k where positive constants.

D.3 Sample Problem 3

Consider the boundary value problem (Fig. D.1), with domain $\Omega = (0, L)$:

$$\boxed{\begin{array}{l} \dfrac{d}{dx}\left(E\dfrac{du}{dx}\right) + k_1\dfrac{du}{dx} + k_2u = cos(14\pi x/L) \\ \\ E\dfrac{du}{dx}(x = 0) = c_1, \\ \\ u(x = L) = c_2, \end{array}} \tag{D.7}$$

where $E > 0$, k_1, and k_2 are constants.

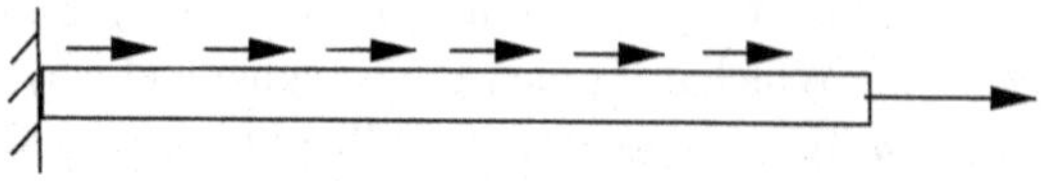

Fig. D.1 A 1D structure

- (a) Derive the weak form (step-by-step, without the penalty approach).
- (b) If N equally sized **linear** elements were used, set up the matrix system of equations that one would need to solve. Draw the shape functions on the master element. Explicitly determine the element by element contributions (leave in integral form). In other words, derive the linear algebraic system

$$[K]\{a\} = \{R\}. \tag{D.8}$$

  Explicitly write $[K]^e$, $\{a\}$ and $\{R\}^e$. Use isoparametric mappings and make all calculations over the *master element*. Do not evaluate the integrals.
- (c) If N equally sized **quadratic** elements were used, set up the matrix system of equations that one would need to solve. Draw the shape functions on the master element. Explicitly determine the element-by-element contributions (leave in integral form). In other words, derive the linear algebraic system

$$[K]\{a\} = \{R\}. \tag{D.9}$$

  Explicitly write $[K]$, $\{a\}$, and $\{R\}$. Use isoparametric mappings and make all calculations over the *master element*. Do not evaluate the integrals.
- (d) If N equally sized **cubic** elements were used, set up the matrix system of equations that one would need to solve. Draw the shape functions on the master element. Explicitly determine the element by element contributions (leave in integral form). In other words, derive the linear algebraic system

$$[K]\{a\} = \{R\}. \tag{D.10}$$

  Explicitly write $[K]$, $\{a\}$, and $\{R\}$. Use isoparametric mappings and make all calculations over the *master element*. Do not evaluate the integrals.
- (e) How many Gauss points would be needed to evaluate each of the needed integrals using **linear** element shape functions? How many Gauss points would be needed to evaluate each of the needed integrals using **quadratic** element shape functions? How many Gauss points would be needed to evaluate each of the needed integrals, using **cubic** element shape functions?
- (f) Using N **linear** elements, if one used a direct (Gaussian solver) how much would it "cost" to solve? Under what conditions could you use an element-by-element Conjugate Gradient solver? How much would it cost? Explain any expressions that you write down.
- (g) Using N **quadratic** elements, if one used a direct (Gaussian solver) how much would it "cost" to solve? Under what conditions could you use an element-by-element Conjugate Gradient solver? How much would it cost? Explain any expressions that you write down.
- (h) Using N **cubic** elements, if one used a direct (Gaussian solver) how much would it "cost" to solve? Under what conditions could you use an element-by-element Conjugate Gradient solver? How much would it cost? Explain any expressions that you write down.

- (i) Assuming that you know the true solution, u, give an expression for the error in the $H^1(0, L)$ norm.
- (j) For the previous part of this problem, repeat parts (a, b, c, d) with the penalty method to apply the specified (primal/Dirichlet) boundary conditions. It is adequate to simply show the modifications.

D.4 Sample Problem 4

Given the following integral

$$\int_4^{14} \left((x-11)^5 + (x-1)^2 + 12\right) dx, \tag{D.11}$$

using Gaussian quadrature, and assuming that the standard Gauss point weights and locations are given to you, show how you would evaluate it. Indicate how many quadrature points you will need to integrate it exactly.

D.5 Sample Problem 5

Given the following integral

$$I = \int_\Omega (ax^3 + by^5)\, d\Omega, \tag{D.12}$$

where Ω is shown above and using a isoparametric mapping for a quadratic element (Fig. D.2).

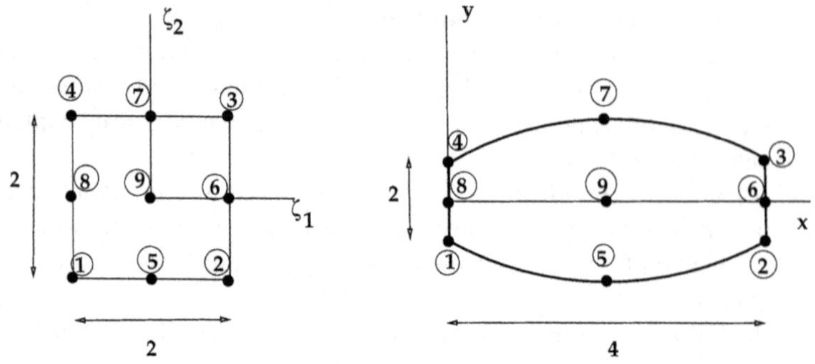

Fig. D.2 A quadratic element

- (a) Derive the quadratic shape functions for the standard 2D (9-node) square master element.
- (b) For the following 2D element, derive the isoparametric mapping for this element shown in the figure.
- (c) Calculate the deformation gradient matrix $\boldsymbol{F}$ for the given mapping and the determinant (Jacobian $J = det\boldsymbol{F}$).
- (d) Set up the integral in the master domain.
- (e) Set up the approximation of this integral with Gaussian quadrature.
- (f) What is the minimum number of Gauss points that are needed in each direction?

D.6 Sample Problem 6

- (a) Carefully, derive the best approximation theorem for $B(u, v) = L(v)$ and support the claim that the FEM solution is "the best possible" (specify in which norm). Include a detailed diagram of the spaces of approximations for u, v, u^h, and v^h.
- (b) What is the (true) potential $\mathcal{J}(u)$ for that problem?
- (c) Explain why the potential is important to monitor in the finite element method for this class of problems.
- (d) How could one could use the potential to determine the constant C in the error estimation expression:

$$||u - u^h||_{E(\Omega)} \leq Ch \tag{D.13}$$

  by using two successively finer meshes. Be very explicit.
- (e) Given

$$\frac{d}{dx}\left(E(\frac{du}{dx})\right) + Ku = f, \tag{D.14}$$

  and boundary conditions $u(0) = c_1$, $\frac{du}{dx}|_{x=L} = c_2$. Derive the weak form:

$$B(u, v) = L(v), \tag{D.15}$$

  and the potential $\mathcal{J}(\boldsymbol{u})$.
- (f) For this system, under what conditions will the operator $\sqrt{B(u, u)}$ violate being a norm?

D.7 Sample Problem 7

Consider the following $(N \times N)$ matrix equation:

$$[K]\{a\} = \{R\}. \tag{D.16}$$

- (a) When can one apply the Conjugate Gradient Method to obtain a solution?
- (b) Explicitly, write down what we are trying to minimize to obtain a solution.
- (c) Derive the ingredients needed for the Conjugate Gradient Method.
- (d) If $[K]$ happened to come from a one-dimensional finite element discretization using linear elements, how much cheaper would it be to solve using a CG element by element solver than a direct Gaussian approach that does not exploit the element-by-element structure.
- (e) Define the condition number of $[K]$. How does your answer in (d) depend on the condition number?
- (f) Given the following matrix:

$$[K] = \begin{bmatrix} 4 & 0 & 1 \\ 0 & 5 & 0 \\ 1 & 0 & 6 \end{bmatrix}, \tag{D.17}$$

  what is the condition number?
- (g) Given the previous matrix and $R = (1, 1, 1)^T$ and $a^{i=0} = (2, 2, 2)$, perform two complete CG iterations.
- (h) Now use a diagonal preconditioner (like in class), and repeat part (g).
- (i) What is the condition number of the new system?
- Repeat for:

$$[K] = \begin{bmatrix} 3 & 2 & 0 \\ 2 & 3 & 0 \\ 0 & 0 & 5 \end{bmatrix}. \tag{D.18}$$

- Repeat for:

$$[K] = \begin{bmatrix} 4 & -1 & 0 \\ -1 & 4 & 0 \\ 0 & 0 & 8 \end{bmatrix}. \tag{D.19}$$

- Repeat for:

$$[K] = \begin{bmatrix} 5 & 0 & 3 \\ 0 & 4 & 0 \\ 3 & 0 & 2 \end{bmatrix}. \tag{D.20}$$

- Repeat for:

$$[K] = \begin{bmatrix} 3 & 0 & 5 \\ 0 & 4 & 0 \\ 5 & 0 & 10 \end{bmatrix}. \tag{D.21}$$

D.8 Sample Problem 8

- (a) Explain the concept of an isoparametric map in 1D, 2D, and 3D and indicate the mathematical condition one must avoid to have "good" elements.
- (b) Using an isoparametric map, construct the mapping function for each of the 2D elements on in Fig. D.3.
- (c) What, if anything, is wrong with the following elements? Explicitly show what you mean mathematically.
- (d) Determine the Jacobian of the mapping for each element.
- (e) Repeat for the 2D elements in Fig. D.4.

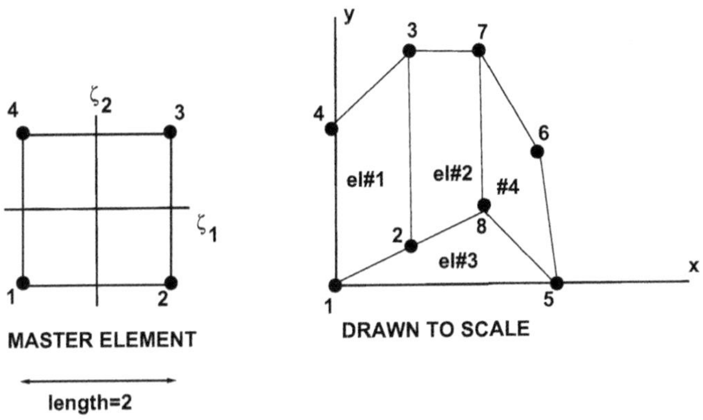

Fig. D.3 Element group #1

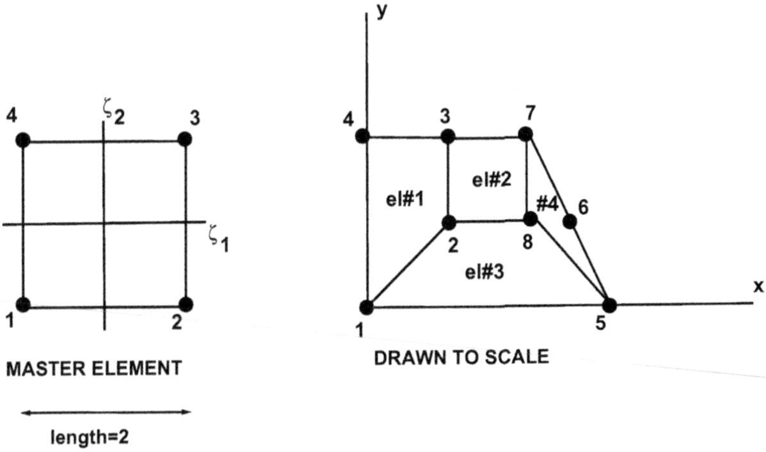

Fig. D.4 Element group #2

D.9 Sample Problem 9

- (a) Give the proper connectivity functions for the mesh in Fig. D.5. Use the standard counterclockwise for the local numbering on the master element (Tables D.1, D.2, and D.3).
- (b) Using the standard counterclockwise local numbering for the master element, is there anything wrong with the following connectivity? If so, what problem will occur?

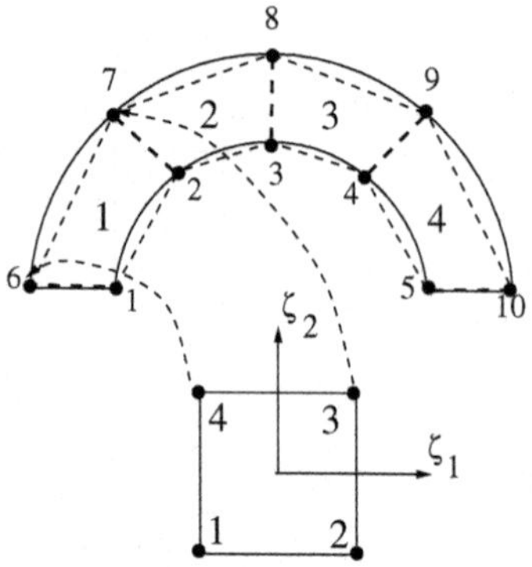

Fig. D.5 Hypothetical mapping

Table D.1 Above: local/global numbers for elements (YOU HAVE 4 ELEMENTS)

Local number	EL # 1-GLOB	EL # 2-GLOB	EL # 3-GLOB	EL # 4-GLOB
1	3	7	1	8
2	4	8	5	5
3	1	2	8	6
4	2	3	2	7

Table D.2 Above: local/global numbers for elements for part (a)

Local	EL # 1-GLOB	EL # 2-GLOB	EL # 3-GLOB	EL # 4-GLOB
1				
2				
3				
4				

Table D.3 Above: local/global numbers for elements for part (b)

Local	EL # 1-GLOB	EL # 2-GLOB	EL # 3-GLOB	EL # 4-GLOB
1	6	7	8	4
2	1	2	3	5
3	2	3	9	10
4	7	8	4	9

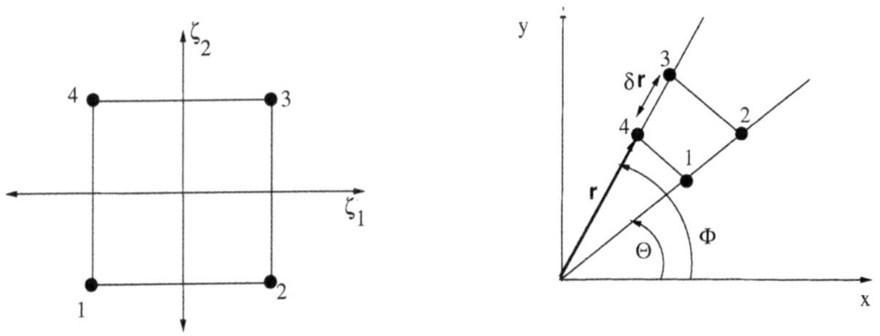

Fig. D.6 First hypothetical mapping

D.10 Sample Problem 10

- (a) (5 points) Derive the linear shape (basis) functions for the standard 2D square master element (Fig. D.6).
- (b) (5 points) For the following 2D element (Fig. D.7), derive the isoparametric map.
- (c) (5 points) What, if anything, is wrong with the following elements? Explicitly show what you mean mathematically. In the second element nodes 1 and 4 coincide.

D.11 Sample Problem 11

- (a) Derive the quadratic shape functions for the standard 2D (9-node) square master element in Fig. D.8.
- (b) For the following 2D element, derive the isoparametric mapping for this element shown in the figure.
- (c) Using the isoparametric mapping, what is the Jacobian of the mapping for this element?

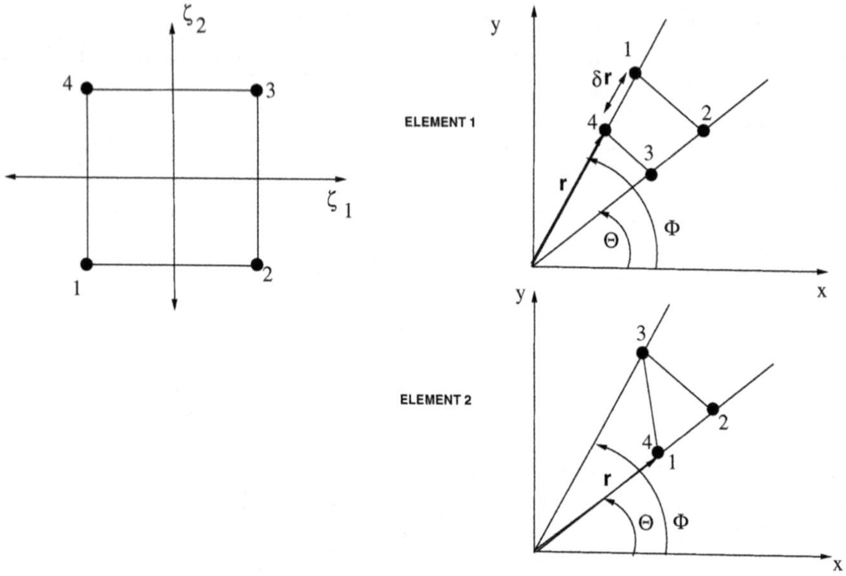

Fig. D.7 Second hypothetical mapping

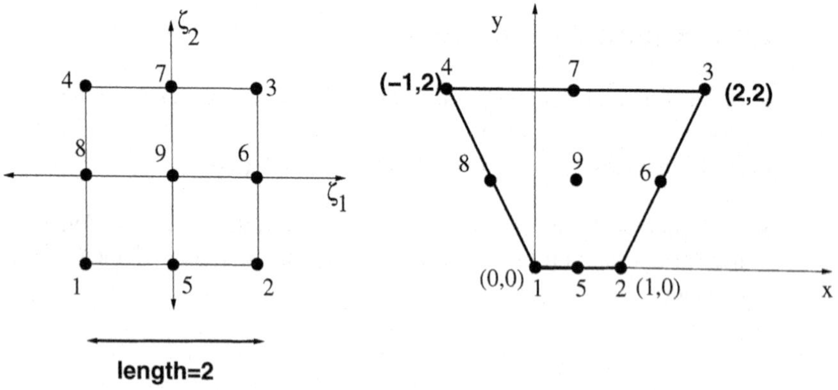

Fig. D.8 Quadratic 2D element

D.12 Sample Problem 12

With the standard isoparametric mapping for a trilinear cube element:

$$x(\zeta_1, \zeta_2, \zeta_3) = \sum_{i=1}^{8} X_i \phi_i(\zeta_1, \zeta_2, \zeta_3) \tag{D.22}$$

and

$$y(\zeta_1, \zeta_2, \zeta_3) = \sum_{i=1}^{8} Y_i \phi_i(\zeta_1, \zeta_2, \zeta_3) \tag{D.23}$$

and

$$z(\zeta_1, \zeta_2, \zeta_3) = \sum_{i=1}^{8} Z_i \phi_i(\zeta_1, \zeta_2, \zeta_3) \tag{D.24}$$

where ϕ_i are the standard shape functions, e.g., $\phi_i = \frac{1}{8}(1 \pm \zeta_1)(1 \pm \zeta_2)(1 \pm \zeta_3)$ and X_i, Y_i, Z_i are the nodal positions. Describe in detail how to solve for $\zeta_1, \zeta_2, \zeta_3$ with Newton's method.

D.13 Sample Problem 13

Consider an ODE given by

$$\dot{u} + 4e^{u+1} = sin(\omega t) \tag{D.25}$$

where $C > 0$ and ω are a constant and $u(0) = k$.

- (a) Set up the solution process using an explicit time-stepping scheme.
- (b) Set up the solution process using an implicit time-stepping scheme and Newton's method.
- (c) What are the general differences (pros/cons) between explicit and implicit methods. Support claims with examples.
- (d) Repeat for:

$$\dot{u} + 5(u+1)^{11} = cos(\omega t) \tag{D.26}$$

  where $C > 0$ and ω are a constant and $u(0) = k$.
- (e) Repeat for:

$$\dot{u} - C(u+5)^5 = 0 \tag{D.27}$$

  where $C > 0$ is a constant and $u(0) = k$.
- (f) Repeat for:

$$\dot{u} - C(cosu + 1)^4 = 0 \tag{D.28}$$

  where $C > 0$ is a constant and $u(0) = k$.

D.14 Sample Problem 14

You are given an $(L \times L \times L)$ block (Fig. D.9) with *scalar* diffusivity $D(x, y, z) > 0$, where the concentration is governed by

$$\dot{c} = \nabla \cdot (D \nabla c) + f - \tau c, \tag{D.29}$$

where $f = f(x, y, z)$ is given data (sources), reaction coefficient $\tau = \tau(x, y, z)$ and with the initial condition $c(t = 0, x, y, z) = c_0(x, y, z)$. It is externally flux loaded on a portion of its surface Γ_g

$$(D \nabla c) \cdot \boldsymbol{n} = g, \tag{D.30}$$

where g is given and $\boldsymbol{n}$ is the outward surface normal, while it has as specified concentration on a portion of its surface Γ_c

$$c = c_o. \tag{D.31}$$

Note: The union of Γ_c and Γ_g comprises the entire boundary of the body.

- (a) Now develop a weak form for the problem, providing all the steps and assumptions necessary. Derive it with and without the penalty method. Explain the differences.
- (b) Use a finite difference approximation in time, and give the modified weak form for the implicit method with and without the penalty method. Explain the differences.
- (c) Substitute the basis functions into the weak form, and use the penalty method to apply concentration boundary conditions. Be very explicit in indicating what integral terms contribute to which "matrix" and "vector" terms in the system that you would eventually have to solve (**Just set it up**) with and without the penalty method.

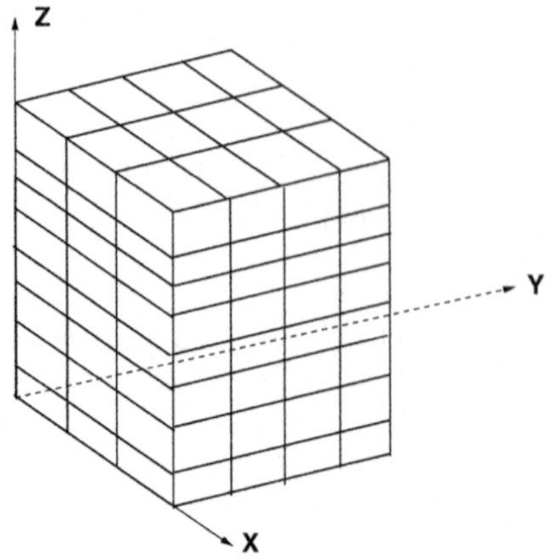

Fig. D.9 A block experiencing diffusion

- (d) Now use a finite difference approximation in time, and give the modified weak form for the explicit method.
- (e) Substitute the basis functions into the weak form, and use the penalty method to apply concentration boundary conditions. Be very explicit in indicating what integral terms contribute to which "matrix" and "vector" terms in the system that you would eventually have to solve (**Just set it up**) with and without the penalty method.

D.15 Sample Problem 15

Consider the standard heat conduction problem in two dimensions (Fig. D.10):

$$\nabla \cdot \mathbb{K} \cdot \nabla T + f = 0 \quad \text{(D.32)}$$

that you are to solve on the following two-dimensional mesh:

The Dirichlet boundary condition has a temperature $T = T^*$, and the Neumann (flux) conditions are applied to every other surface ($g = (\mathbb{K} \cdot \nabla T) \cdot \boldsymbol{n}$).

- (a) Derive the weak form of the problem, imposing the Dirichlet boundary conditions directly. Be sure to provide details on the spaces that the functions live in.
- (b) Sketch the matrix system to be solved, showing how the boundary conditions are applied. Just put X's to denote nonzero entries.
- (c) Derive the weak form of the problem, imposing the Dirichlet boundary conditions using the penalty method. Be sure to detail from the spaces that the functions live in. You only need to state the additional terms added to the form from part (a).
- (d) Sketch the matrix system to be solved, showing how the boundary conditions are applied. State and illustrate the changes caused by the penalty terms.

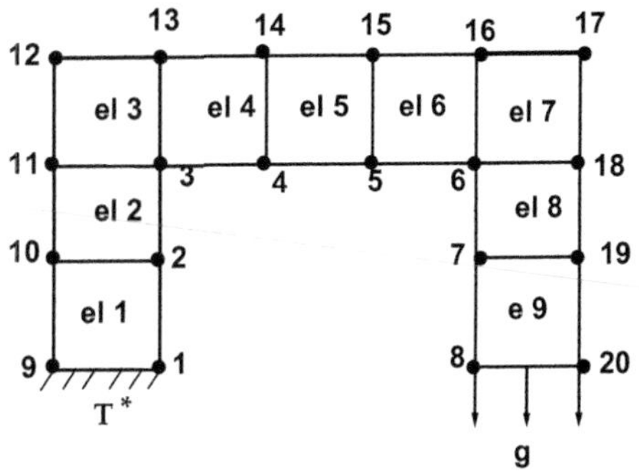

Fig. D.10 An arch experiencing heat transfer

- (e) Suppose you are using the Conjugate Gradient Method in both cases. How does solving the matrix system change when using the penalty method? State two differences.

D.16 Sample Problem 16

Consider the following linear elasticity problem, called “problem 1”:

$$problem\ \#1: \ \nabla \cdot (\mathbb{E} : \nabla \boldsymbol{u}) + \boldsymbol{f} = \boldsymbol{0}, \tag{D.33}$$

where $\boldsymbol{f}$ is a given source term and the following heat conduction problem, called “problem 2”:

$$problem\ \#2: \ \nabla \cdot (\mathbb{K} \cdot \nabla T) + w = 0 \tag{D.34}$$

where w is a given source term.

For each problem, consider the three-dimensional block where we use **linear** brick elements resulting in 3N-1 elements in the x-direction, 3N-1 elements in the y-direction, and 3N-1 elements in the z-direction, where N is given.

- (a) How much does it cost if we use a regular Gaussian solver for problem 1 and problem 2?
- (b) How much does it cost if we use a regular CG solver for problem 1 and problem 2?
- (c) How much does it cost if we use a element-by-element CG solver for problem 1 and problem 2?

Now consider the same problem with the same finite element mesh, but now the mesh is broken up using a $5 \times 5 \times 5$ subdomain decompositions and distributed across 125 processors:

- (d) Draw a picture of the system and the decomposition.
- (e) How much does it cost if we use the domain decomposition process and a regular Gaussian solver in each subdomain for problem 1 and problem 2?

D.17 Sample Problem 17

Given:

$$\frac{d}{dx}\left(A\left(\frac{du}{dx}\right)^{p}\right) + u^{q} + f = 0, \tag{D.35}$$

$u(0) = c_1$, $\frac{du}{dx}|_{x=L} = c_2$

- (a) Derive the weak form.
- (b) Derive linearized weak form for the system in part (a) around $u = u_o$.
- (c) If one were to attempt to apply the Conjugate Gradient Method to solve the linearized system, for what combination of A, u_o, p, and q will the CG method fail?

Appendix E
Selected Computer Projects

E

The projects in this chapter are selected from exams given over the last 15 years UC Berkeley.

E.1 Assignment Format

- All assignments must be typed-nothing handwritten.
- Be concise-shorter is better-provided you do not delete essential information.
- You are encouraged to talk and work with one another. Please see me for any problems, theoretical, coding etc.
- Introduction to the problem: explain it to a layman.
- Objectives: what are the goals?
- Your procedure: Brief explanation, flow-charts, difficulties, assumptions, etc.
- Findings: figures, plots and tables. Make sure they are readable.
- Observations and discussion: some interpretation and insight into the results.
- Appendix: the messy stuff like your code or raw data.

E.2 Sample Project 1: The Basics of FEM

- Solve the following boundary value problem, with domain $\Omega = (0, L)$, analytically:

T. I. Zohdi, *A Finite Element Primer for Beginners*, The Basics,
https://doi.org/10.1007/978-3-319-70428-9

$$\boxed{\begin{aligned} &\frac{d}{dx}\left(E\frac{du}{dx}\right) = k^2 sin(\tfrac{2\pi kx}{L}) \\ &E = given\ constant = 0.1 \\ &k = given\ constant \\ &L = 1 \\ &u(0) = \Delta_1 = given\ constant = 0 \\ &u(L) = \Delta_2 = given\ constant = 1 \end{aligned}} \tag{E.1}$$

- Now solve this with the finite element method using linear equal-sized elements. In order to achieve,

$$\boxed{\begin{aligned} &e^N \stackrel{\text{def}}{=} \frac{||u-u^N||_{E(\Omega)}}{||u||_{E(\Omega)}} \leq TOL = 0.05, \\ &||u||_{E(\Omega)} \stackrel{\text{def}}{=} \sqrt{\int_\Omega \frac{du}{dx} E \frac{du}{dx}\, dx} \end{aligned}} \tag{E.2}$$

How many finite elements (N) are needed for

$$\boxed{\begin{aligned} &k = 1 \Rightarrow N = ? \\ &k = 2 \Rightarrow N = ? \\ &k = 4 \Rightarrow N = ? \\ &k = 8 \Rightarrow N = ? \\ &k = 16 \Rightarrow N = ? \\ &k = 32 \Rightarrow N = ? \end{aligned}} \tag{E.3}$$

You should set up a general matrix equation and solve it using Gaussian elimination. Later we will use other types of more efficient solvers. Plot the numerical solutions for $N = 2, 4, 8, 16, \ldots$, for each k, along with the exact solution. Also make a plot of the e^N for each k.

Remarks: You should write a general one-dimensional code where you specify the number of elements. Your code should partition the domain automatically. However, if you want to make the code more general (for future assignments), you should put in the following features:

- element endpoint locations (different sized elements),
- the possibility for different material values for each element ($E(x)$).

E.3 Sample Project 2: Higher-Order Elements

- Consider the following boundary value problem, with domain $\Omega = (0, L)$:

$$\boxed{\begin{aligned} &\frac{d}{dx}\left(E\frac{du}{dx}\right) = xk^3 cos\left(\frac{2\pi k x}{L}\right)\\ &E = 0.2\\ &k = 12\\ &L = 1\\ &u(0) = \Delta_1 = given\ constant = 3\\ &u(L) = \Delta_2 = given\ constant = -1 \end{aligned}} \tag{E.4}$$

- Solve this with the finite element method using order p equal-sized elements. In order to achieve

$$\boxed{\begin{aligned} &e^N \stackrel{\text{def}}{=} \frac{||u-u^N||_{E(\Omega)}}{||u||_{E(\Omega)}} \leq TOL = 0.04,\\ &||u||_{E(\Omega)} \stackrel{\text{def}}{=} \sqrt{\int_{\Omega} \frac{du}{dx} E \frac{du}{dx}\, dx} \end{aligned}} \tag{E.5}$$

how many finite elements (N) are needed for

$$\boxed{\begin{aligned} &p = 1 \Rightarrow N = ?\\ &p = 2 \Rightarrow N = ?\\ &p = 3 \Rightarrow N = ? \end{aligned}} \tag{E.6}$$

- Plot the numerical solutions for several values of N, for each p, along with the exact solution.
- Plot e^N as a function of the element size h for each p.
- Plot e^N as a function of the number of degrees of freedom for each p.
- Determine the relationship between the error and the element size for each p.
- Note: Please be careful with the quadrature order...you will need higher order Gauss rules for quadratic and cubic elements.

E.4 Sample Project 3: Potential and Efficient Solution Techniques

- Solve the following boundary value problem, with domain $\Omega = (0, L)$, analytically:

$$
\begin{aligned}
&\frac{d}{dx}\left(E(x)\frac{du}{dx}\right) = xk^3 cos(\frac{2\pi kx}{L}) \\
&E(x) = 10\ different\ segments\ (see\ below) \\
&k = 12,\ L = 1,\ u(0) = -0.3,\ u(L) = 0.7
\end{aligned}
\tag{E.7}
$$

For E

$$
\begin{aligned}
&FOR\ 0.0 < x < 0.1\ E = 2.5 \\
&FOR\ 0.1 < x < 0.2\ E = 1.0 \\
&FOR\ 0.2 < x < 0.3\ E = 1.75 \\
&FOR\ 0.3 < x < 0.4\ E = 1.25 \\
&FOR\ 0.4 < x < 0.5\ E = 2.75 \\
&FOR\ 0.5 < x < 0.6\ E = 3.75 \\
&FOR\ 0.6 < x < 0.7\ E = 2.25 \\
&FOR\ 0.7 < x < 0.8\ E = 0.75 \\
&FOR\ 0.8 < x < 0.9\ E = 2.0 \\
&FOR\ 0.9 < x < 1.0\ E = 1.0
\end{aligned}
\tag{E.8}
$$

- Solve this with the finite element method using linear equal-sized elements. Use 100, 1000, and 10000 elements. You are to write a preconditioned Conjugate Gradient solver. Use the diagonal preconditioning given in the notes. The data storage is to be done element by element (symmetric), and the matrix vector multiplication is to be done element by element.
- You are to plot the solution (nodal values) for each N.
- You are to plot

$$
\begin{aligned}
&e^N \stackrel{\text{def}}{=} \frac{||u - u^N||_{E(\Omega)}}{||u||_{E(\Omega)}}, \\
&||u||_{E(\Omega)} \stackrel{\text{def}}{=} \sqrt{\int_\Omega \frac{du}{dx} E \frac{du}{dx}\, dx},
\end{aligned}
\tag{E.9}
$$

for each N.

- You are to plot

$$Potential\ energy = \mathcal{J}(u^N) \tag{E.10}$$

for each N.
- You are to plot the number of PCG solver iterations for each N for a stopping tolerance of 0.000001.
- Use a Gauss integration rule of level 5.
- Check your Conjugate Gradient generated results against a regular Gaussian solver, for example the one available in MATLAB.

E.5 Sample Project 4: Error Estimation and Adaptive Meshing Using the Exact Solution as a Test

- Consider the boundary value problem $\frac{d}{dx}\left(E\frac{du}{dx}\right) = f(x)$, $E = 1$, with domain $\Omega = (0, L)$, $L = 1$, and solution $u(x) = \cos(10\pi x^5)$.
- Compute the finite element solution u^N to this problem using linear equal-sized elements. Determine how many elements are needed in order to achieve

$$e^N \stackrel{\text{def}}{=} \frac{||u - u^N||_{E(\Omega)}}{||u||_{E(\Omega)}} \leq TOL = 0.05\,,$$

$$||u||_{E(\Omega)} \stackrel{\text{def}}{=} \sqrt{\int_{\Omega} \frac{du}{dx} E \frac{du}{dx} dx}$$

- Plot I versus A_I, where

$$A_I^2 \stackrel{\text{def}}{=} \frac{\frac{1}{h_I}||u - u^N||_{E(\Omega_I)}^2}{\frac{1}{L}||u||_{E(\Omega)}^2}\,.$$

Here I is the element index, h_I is the length of element I, and

$$||u||_{E(\Omega_I)}^2 \stackrel{\text{def}}{=} \int_{\Omega_I} \frac{du}{dx} E \frac{du}{dx} dx\,.$$

- Modify your code from HW 1 so that it can automatically refine the mesh the following criterion:
- Refine the mesh (by dividing elements into two) until $A_I < TOL_E$ for all I. Use this criterion to refine your mesh, starting with $N = 20$ equal-sized elements:
  - Determine how many elements are needed to achieve $A_I < TOL_E = 0.05$ for all I.
  - Plot the final solution, together with the exact solution.

- Tabulate the final number of elements that fall into each of the initial 20 elements.
- Plot X_I versus A_I for the final solution (X_I = position of node I).

E.6 Sample Project 5: 3D Formulations for Elasticity

You are given a tubular multiphase structure (Fig. E.1) with an elasticity of $\mathbb{E}(x, y, z)$, and with dimensions shown in the figure. It is clamped on one end and externally traction loaded everywhere else, including on the interior surface. The small deformation of the body is governed by (strong form):

$$\nabla \cdot (\mathbb{E} : \nabla \boldsymbol{u}) + \rho \boldsymbol{b} = \boldsymbol{0} \tag{E.11}$$

where $\mathbb{E}$ and ρ are spatially variable and where $\boldsymbol{b} = \boldsymbol{b}(x, y, z)$ is given data.

- Develop a weak form, providing all the steps and assumptions necessary. Carefully define the spaces of approximation.
- Develop a finite element weak form. Carefully define the spaces of approximation.
- Develop a finite element weak statement using the penalty method. Carefully define the spaces of approximation.
- Derive the equations for element stiffness matrices (be explicit) and load vectors. Thereafter, describe how the global stiffness matrix and load vector are generated, using the penalty method. Use trilinear subspatial approximations. There are different kinds of loading on the surfaces, so be very explicit as to what each of the individual stiffness matrices and right-hand-side vectors look like, as well as a generic element that is not on the surface.
- Write a mesh generator. Explicitly explain how it works and, in particular, the connectivity function. Use N_t elements in the thickness direction, N_c elements in

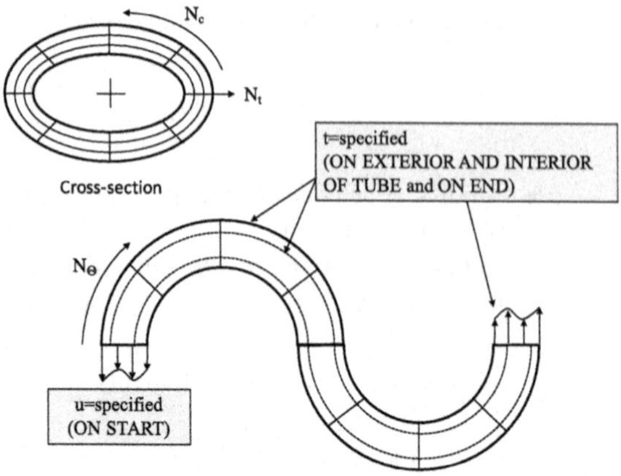

Fig. E.1 3D structure

the circumferential direction, and N_θ elements in the θ direction for each semi-circular portion. For the given figure $N_t = 3$, $N_c = 4$, and $N_\theta = 8$ *generate the mesh-show it.*

- If one were to use a Conjugate Gradient solver, theoretically how many operation counts would be needed to solve this problem for a mesh of N_t elements in the thickness direction, N_c elements in the circumferential direction, and N_θ elements in the θ direction

E.7 Sample Project 6: Implementation of the Finite Element Method in 2D

- Solve the following boundary value problem, on an arch-shaped domain, using the finite element method:

$$\nabla \cdot (K \nabla T) + f = 0,$$

$$T = T_0 \text{ along } \theta = \pi$$

$$-K\nabla T \cdot \mathbf{n} = q_0(r) \text{ along } \theta = 0$$

$$-K\nabla T \cdot \mathbf{n} = 0 \text{ along } r = r_i, r_o$$

$$K = K_1 \text{ for } ||\mathbf{x} - \mathbf{x}_c|| \le r_c$$

$$K = K_2 \text{ for } ||\mathbf{x} - \mathbf{x}_c|| > r_c$$

These equations describe a thermal physics problem of the two-phase structure that is shown in Fig. E.2.

- You are to generate a mesh of the domain for $r_i = 3$ and $r_o = 4$. Use $N_r \times N_\theta$ quadrilateral elements. For example, in Fig. E.3, $N_r = 3$ and $N_\theta = 12$. Write a finite element code to solve this problem with bilinear shape functions. For elements that have material discontinuities use a 5×5 Gaussian integration rule, otherwise, use a 2×2 rule.

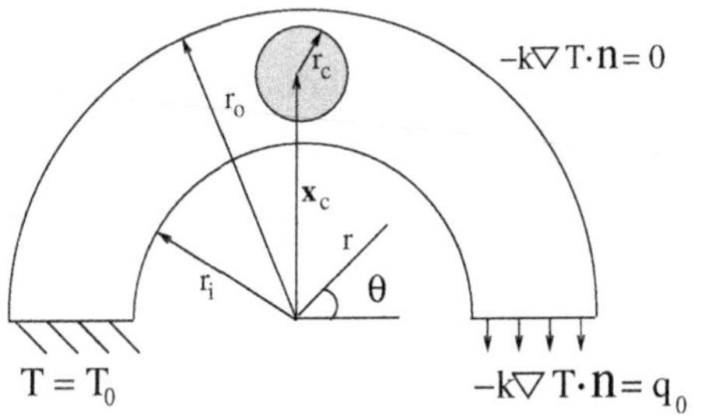

Fig. E.2 An arch

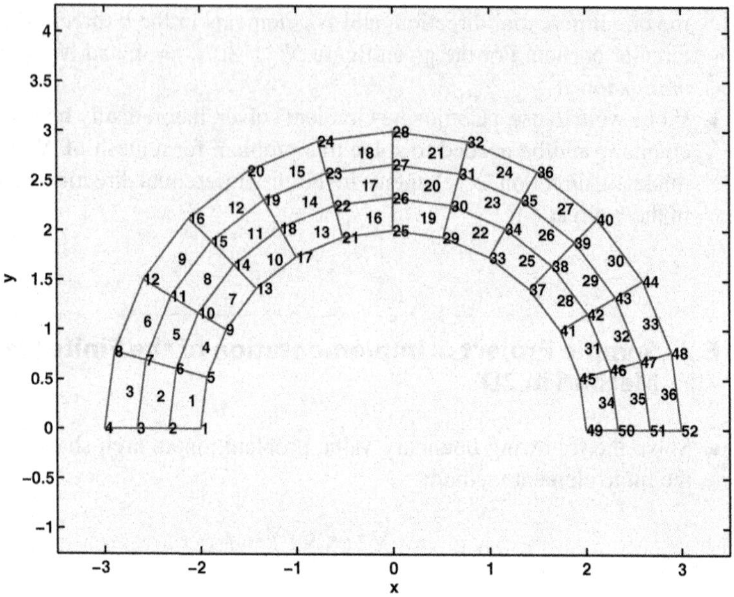

Fig. E.3 The proposed mesh for the arch

- Solve the problem both by strictly enforcing the boundary conditions and by using the penalty method on Γ_T (the part of the boundary where the temperature is prescribed). Explain how your choice of penalty parameter affects the results.
- To verify that your code works properly, solve this with $k_1 = k_2 = 1$, $T_0 = 100$, $q_0(r) = \frac{40}{r}$, and $f(r, \theta) = \frac{80}{r^2} sin(2\theta)$. Determine the exact solution for this problem (Hint: the solution is independent of r). Include a plot of your solution for $N_r = 10$ and $N_\theta = 80$.
- Solve the problem with $k_1 = 10^{-3}$, $k_2 = 1$, $T_0 = 110$, $q_0(r) = \frac{20}{r}$, $f(r, \theta) = \frac{40}{r^2} sin(2\theta)$, $r_c = 0.40$, and $\mathbf{x}_c$ is given by $(r = 3.5, \theta = \pi/2)$. Include a plot of your solution for $N_r = 50$ and $N_\theta = 400$.

E.8 Sample Project 7: Time-Dependent Problems

Part 1: formulation

You are given a two-phase (two material) structure (Fig. E.4), comprised of a two semicircular rings, with *scalar* diffusivity $D(x, y, z)$ and with dimensions shown in the figure. It is externally flux loaded on a portion of the top surface and one end surface, while the other end surface has as specified concentration. All other

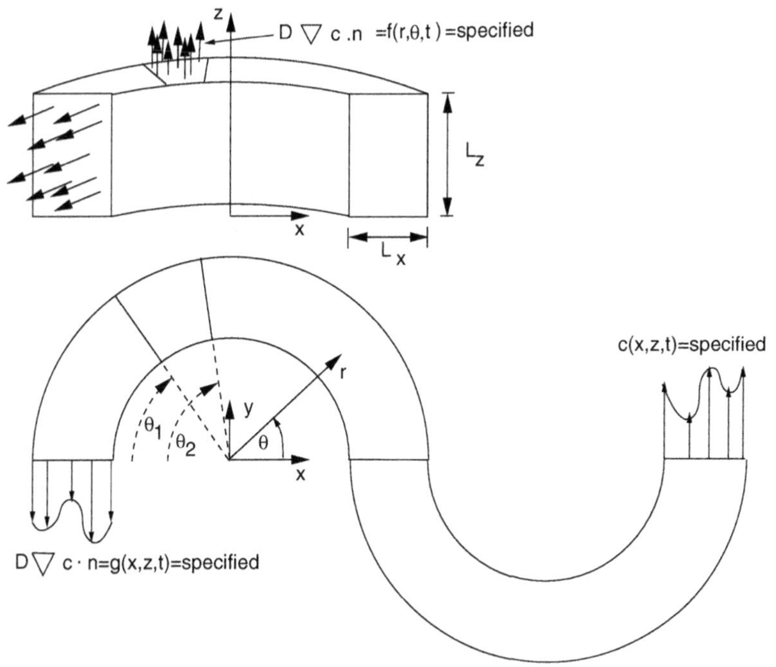

Fig. E.4 3D structure

surfaces are free of any "loading" (flux-free). The physics of the body is described by a simplified version of the diffusion-reaction equation, which in strong form is:

$$\nabla \cdot (D\nabla c) - \tau c + f = 0, \tag{E.12}$$

where D is a nonconstant, positive scalar function and where $f = f(x, y, z)$ is given data (sources).

- Now develop a weak form for the statement, providing all the steps and assumptions necessary.
- Develop a finite element weak statement. Carefully define the spaces of approximation.
- Develop a finite element weak statement using the penalty method. Carefully define the spaces of approximation.
- Derive explicit equations for element stiffness matrices and load vectors. Thereafter, describe how the global stiffness matrix and load vector are generated, using the penalty method. Use trilinear subspatial approximations. There are different kinds of loading on the surfaces, so be very explicit as to what each of the individual stiffness matrices and right-hand-side vectors look like, as well as a generic

element that is not on the surface. Also, pay attention to the fact that elements may or may not have discontinuities when using Gaussian integration (consider both cases).

- Using your mesh generator, modify it to handle this new problem. Explicitly explain how it works and, in particular, the connectivity function. Use N_t elements in the thickness direction, N_c elements in the circumferential direction, and N_θ elements in the θ direction for each semicircular portion.
- If one were to use a Conjugate Gradient solver, theoretically how many operation counts would be needed to solve this problem for a mesh of N_t elements in the thickness direction, N_c elements in the circumferential direction, and N_θ elements in the θ direction.
- Now consider the time-transient case. The body has the same boundary conditions as before, with the initial condition that $c(t = 0, x, y, z) = c_0(x, y, z)$. The governing equation is

$$\nabla \cdot (D\nabla c) - \tau c + f = \dot{c}, \tag{E.13}$$

  Develop a finite element weak statement. Carefully define the spaces of approximation. Use the finite difference approximation that we have used this semester for the time-dependent term.
- Finally, given that this is a three-dimensional problem, with heterogeneous coefficients, it is most likely you will need a large number of elements to solve it. Suppose that your machine has only enough memory to allow you to solve a wedge (sector) of $0 \le \theta \le \theta^* << \pi$ degrees, but that you need to solve the entire $0 \le \theta \le \pi$ problem. How would you break the problem and solve it? Give an overall algorithm.

Part 2: Implementation in 1D

- Solve the following boundary value problem ($L = 1$)

$$\boxed{\begin{array}{l}
\frac{\partial c}{\partial t} = \frac{\partial}{\partial x}\left(D\frac{\partial c}{\partial x}\right) - \tau c \\[2ex]
D(x) = 10\ different\ segments\ (see\ below) \\[2ex]
\tau(x) = 10\ different\ segments\ (see\ below) \\[2ex]
c(x = 0, t) = 0.5 \\[2ex]
D\frac{\partial c}{\partial x}(x = L, t) = 5 \times 10^{-6} \\[2ex]
c(x, t = 0) = 0.5 \quad 0 < x < L
\end{array}} \tag{E.14}$$

with 100 elements ($\delta x = 0.01$) and set the total amount of time to be $T = 6500\,\text{s}$. Use an implicit (Backward Euler) time-stepping scheme. Solve with the following time-step sizes: (I) $\delta t = \frac{T}{100}$, (II)$\delta t = \frac{T}{1000}$, and (III)$\delta t = \frac{T}{10000}$ with

$$\begin{array}{|l|}\hline For\ 0.0 < x < 0.1\ D = 2.4 \times 10^{-6} \\ For\ 0.1 < x < 0.2\ D = 2.0 \times 10^{-6} \\ For\ 0.2 < x < 0.3\ D = 1.5 \times 10^{-6} \\ For\ 0.3 < x < 0.4\ D = 0.6 \times 10^{-6} \\ For\ 0.4 < x < 0.5\ D = 1.3 \times 10^{-6} \\ For\ 0.5 < x < 0.6\ D = 0.14 \times 10^{-6} \\ For\ 0.6 < x < 0.7\ D = 1.1 \times 10^{-6} \\ For\ 0.7 < x < 0.8\ D = 2.2 \times 10^{-6} \\ For\ 0.8 < x < 0.9\ D = 2.0 \times 10^{-6} \\ For\ 0.9 < x < 1.0\ D = 1.5 \times 10^{-6} \\ \hline \end{array} \tag{E.15}$$

$$\begin{array}{|l|}\hline For\ 0.0 < x < 0.1\ \tau = 1.2 \times 10^{-3} \\ For\ 0.1 < x < 0.2\ \tau = 0.8 \times 10^{-3} \\ For\ 0.2 < x < 0.3\ \tau = 0.3 \times 10^{-3} \\ For\ 0.3 < x < 0.4\ \tau = 1.4 \times 10^{-3} \\ For\ 0.4 < x < 0.5\ \tau = 1.15 \times 10^{-3} \\ For\ 0.5 < x < 0.6\ \tau = 0.75 \times 10^{-3} \\ For\ 0.6 < x < 0.7\ \tau = 0.35 \times 10^{-3} \\ For\ 0.7 < x < 0.8\ \tau = 0.85 \times 10^{-3} \\ For\ 0.8 < x < 0.9\ \tau = 1.25 \times 10^{-3} \\ For\ 0.9 < x < 1.0\ \tau = 2.0 \times 10^{-3} \\ \hline \end{array} \tag{E.16}$$

- Clearly show the evolution of solution with time, together with the final time solution. Also, comment on the effect of time-step size on the solution.

References

1. Malvern, L. (1968). *Introduction to the mechanics of a continuous medium*. New York: Prentice Hall.
2. Gurtin, M. (1981). *An introduction to continuum mechanics*. New York: Academic Press.
3. Chandrasekharaiah, D. S., & Debnath, L. (1994). *Continuum mechanics*. San Diego: Academic Press.

Printed in the United States
By Bookmasters